Shortwave Listening Guidebook

Shortwave Listening Guidebook

Harry L. Helms

A DX/SWL Press Book
HighText Publications, Inc.
San Diego

Printed in the United States of America

Cover and interior design: Frank Soley
Developmental editing: Carol Lewis

ISBN 1-878707-02-7

Contents

Preface

Listening to shortwave radio may seem like an anachronism in an age when average citizens routinely receive television programs using satellite dishes in their backyards and telephone calls are routed through glass fibers using light. By all logic, shortwave radio should have gone the way of telegraphy over land wires. But the hobby of shortwave listening is bigger—and livelier—today than ever before. Why is this?

Everyone who listens to shortwave radio probably has a different explanation. For me, it's because a shortwave radio is a magic box.

Over twenty-five years ago I was a junior high student. I became curious as to why I could hear distant stations on an AM radio at night but not in the daytime. That puzzle led me into library stacks looking for an explanation in the various books on radio. In those books, I found out about some other types of radio signals, known collectively as "shortwave," which could be heard all over the world with relative ease. Intrigued, I pestered my parents for a shortwave radio. On my next birthday, my wish came true as I was presented with a simple shortwave receiver manufactured by a now-defunct company known as Hallicrafters.

It was then that I discovered the magic inside a shortwave radio.

All I had to do was twist the dial and I could hear programs—in English, no less!—from such countries as the Soviet Union, Great Britain, West Germany, Spain, Australia, and Japan. Another twist of the dial and I heard programs in languages I never heard spoken before, along with strange music I never knew existed. I could hear ham radio operators talking to each other from all over the country and world. I found many stations sending the dots and dashes of the Morse code; since I didn't know Morse code, they remained a mystery to me. I managed to hear people placing telephone calls from ships at sea along with aircraft in contact with airports. And since this was before the era of communications satellites, I also listened in on many telephone conversations from the United States to Europe.

That shortwave radio became a window on the world for me. Without intending to, I learned much about world politics, customs, cultures, and lifestyles. I found I could automatically name the capital of any country one might mention. I started picking up bits and pieces of various foreign languages and had an endless supply of practice material when I later studied French and Spanish in school. My desire to know more about how my shortwave radio worked pushed me into obtaining my own ham radio license and later into a career as a writer and editor of books on electronics and computing.

Much has changed since I first learned about shortwave radio. There's more to hear today than ever before. Shortwave radios are easier to use, more compact, and a radio costing less than $200 today can run rings around one that cost over $1000 twenty-five years ago. In fact, only one thing has remained constant.

That "box" is still magic.

CHAPTER ONE

What is Shortwave Listening?

IN MANY WAYS, the terms *shortwave listener (SWL)* and *shortwave listening (SWLing)* are misnomers. If you listen to the news from London over the British Broadcasting Corporation (BBC), then you're a shortwave listener. But someone who tries to hear stations from Great Britain on the standard AM broadcasting band may also refer to him- or herself as an "SWL," as might someone who tunes for distant ("DX") signals on the FM broadcasting band or TV channels. Even those who listen to *longwave* frequencies below the AM broadcasting band might refer to themselves as "shortwave" listeners. An "SWL" is a "SWL" even if he or she never actually listens to shortwave!

So what's the difference between an SWL and a normal radio listener or TV viewer? The answer is that SWLs, regardless of the particular frequencies they listen to, are seeking something out of the ordinary. SWLs want to eavesdrop on signals the general public doesn't normally receive. Sometimes the public doesn't listen to such signals because the equipment required is not widely used (as was the case with shortwave radios until recently). In other cases, the signals in question might require special knowledge or skills to catch, such as those from FM and TV stations located several hundreds or even thousands of miles away. Regardless of the frequencies involved, "SWLs" look at the radio and TV spectrum and actively try to find stations and signals that most people aren't aware can be heard or even exist.

(Since the terms "SWL" and "SWLing" are so ambiguous, there have been some efforts to come up with better ones. Some have pushed "world band radio," "world band listening," and "communications monitoring" for SWLing, and "communications enthusiast," "hobby radio listener," and "communications monitor" for SWL. However, old habits die hard, and we're probably stuck with "SWL" and "SWLing" for the foreseeable future.)

SWLs are a diverse group. Many specialize in trying to hear "DX" signals. ("DX" comes from the old radiotelegraph abbreviation for "distance," and refers to stations which are heard only rarely or with extreme difficulty.) Other SWLs are content to listen to a certain group of high-power, easily heard broadcasters such as Radio Moscow, the BBC, Deutsche Welle ("Voice of Germany"), and Radio Japan. In publications for shortwave fans, you'll often see "DXer" used to describe the former group and "SWL" for the latter. But the terms are used interchangeably; it's common for someone to be a DXer some of the time and an SWL the rest. Most DXers have a few major stations and programs they listen to regularly, while most SWLs enjoy occasionally trying to "dig out" a weak DX signal.

Most people have heard of "ham radio operators." The formal term is *amateur radio operator*, and refers to someone who is licensed to *transmit* as well as receive on shortwave frequencies. Such licenses are issued in the United States by the Federal Communications Commission (FCC) when applicants pass an examination. Many hams got interested in amateur radio through SWLing (as was the case with me) and still actively engage in SWLing. The public and media frequently lump together all radio hobbyists—SWLs, hams, even citizens band (CB) operators—together as "hams." But the term properly refers to those licensed to transmit on shortwave.

(By the way, no license or other official sanction is required to own or operate a shortwave—SW—receiver. You can set up a shortwave radio and antenna and listen as much as you like to anything you want. The only exceptions, which we'll discuss later, involve cellular telephone calls and divulging to others certain types of messages not intended for the general public.)

One of the beauties of SWLing is that the hobby becomes whatever you want it to be. If you want, you can try to hear stations from as many different countries as possible and engage in what can be a ferocious competition with other DXers to hear rare or unusual stations. (Hearing a rare station can be only half the battle; many DXers don't rest until they've received a card or letter from the station confirming they indeed heard it!) Other SWLs seem to find DXing a bit vulgar, and content themselves with listening to a few major international broadcasting stations and sending out regular letters commenting on the programs from such stations. There are even monthly publications devoted to shortwave programming; readers discuss favorite or least favorite stations and programs in a manner (and often at a level) remarkably similar to TV soap opera fan magazines. Other SWLs like to listen to communications from barges cruising the Ohio River, to military communications, or to illegally operating "pirate" radio stations scattered throughout the United States. Sometimes it seems the only things SWLs and DXers of different persuasions have in common is a certain ineffable pleasure they all seem to find in their activities.

So what *is* shortwave listening? It's whatever you want it to be, and what you make it. Let's now look at some of the types of signals that you (whether you consider yourself an SWL or DXer) can hear on a shortwave radio.

Shortwave Broadcasters

The first stations you're likely to notice when tuning a shortwave radio are usually broadcast stations. Like your local AM, FM, or TV stations, these are intended for reception by anyone with an SW receiver.

Shortwave broadcasters are grouped into two broad categories. *International* broadcasters direct their programming to listeners outside the country in which they're located. Such programs are usually in the language(s) of the intended target country or area. For example, Radio Moscow broadcasts extensively to the United States and Canada in English but uses Spanish for most of its programs to Latin America. The Voice of America uses over forty languages for its broadcasts. Most international broadcasters are funded and operated by governments; for example, the Voice of America is operated by the United States Information Agency and receives all funding through Congressional appropriations. In some countries, international broadcasting is under the auspices of a public or statutory corporation; Great Britain's BBC is the best-known such example.

One thing you'll notice is that international broadcasters almost never use call letters, such as WCBS or KIRO. Instead, they use names or slogans like "Radio Japan" or "The Voice of Turkey."

Government-operated and -financed international broadcasters tend to be little more than public relations outlets for the country. Don't expect startling insights or critical analysis of the sponsoring country. Some stations are freer of control by the funding government than others (the BBC is especially fortunate in this regard), but the overall impression is that the majority of state-operated international broadcasters are directing their programs more to the funding authorities in their nation than to listeners in other countries. Program content usually reflects what the funding authorities feel is important for overseas listeners to know about the country rather than what overseas listeners might be interested in. An idealized image of the country is often portrayed in broadcasts, and little effort is made to understand the intended foreign audience. The result is that too many international broadcasters are—to be blunt—boring (unless tedious recitations of industrial output

FIGURE 1-1

Radio Sweden is consistently among the most popular stations among shortwave listeners around the world.

figures or reports on obscure local political parties are of interest to you). The problem is often not one of funds; many countries have made large investments in transmitting equipment and antennas to make sure their signals can be easily heard around the world, but don't pay the same attention to programming. This problem is by no means restricted to smaller international broadcasters; one of the worst offenders in this regard is the Voice of America.

However, even dull programming can be useful and informative to the discerning, politically aware listener. The issues and themes present in international programming can reflect how the sponsoring government wishes to be viewed by the outside world. For a long time, Radio Moscow was a reliable indicator of how issues and events in the world were presented to the average Soviet citizen. (With the Gorbachev era, Radio

Moscow has undergone massive changes; these will be described later.) Sometimes, what is *not* said during international broadcasts is more illuminating than what is said. During the last months of the reign of the Shah of Iran, the English broadcasts from the Voice of Iran took the attitude that nothing was really wrong; there's some noise and confusion, the broadcasts went, but nothing to be concerned about. The Voice of Iran maintained this fiction literally up until the day the Shah departed. Then came a dramatic flip-flop as the station became the Voice of the Iranian Revolution, and poured forth invective against the United States. A similar situation took place in Uganda during the final days of Idi Amin. English broadcasts from Kampala repeatedly warned opposition forces to surrender or face certain death. This pretense continued for a couple of days even after Amin had fled the country. Listening to broadcasts in such circumstances has a certain surreal quality!

Some international broadcasters are funded and operated by religious organizations. These stations are frankly evangelical in orientation; their purpose is less to entertain and inform than to convert. One example is Trans World Radio, which operates shortwave stations in such locations as Monte Carlo, Swaziland, the Netherlands Antilles, Cyprus, and Sri Lanka. (Interestingly, religious broadcasters are currently entirely Christian. Despite the influx of petrodollars into Islamic states, they have funded no comparable proselytizing broadcasters.)

Finally, there is a small but growing number of private, commercial international broadcasters in operation. One is Radio Trans Europe in Sines, Portugal. You can't listen to programs from Radio Trans Europe itself, since it doesn't produce its own programming. Instead, the station sells air time to broadcasters and organizations who wish to broadcast on shortwave to Europe but who do not (or cannot) establish their own station in Europe. Thus, Radio Trans Europe has served as a relay for such stations as Radio Japan and Radio Canada International as well as broadcasting programs produced by private organizations

FIGURE 1-2

Alaska's KNLS is an example of an evangelical shortwave broadcasters. It's currently the only shortwave broadcasting station in Alaska.

like Adventist World Radio. Another such station is Africa Number One, located in Gabon, which has relayed programs by Radio Japan and others.

Private commercial shortwave broadcasters have also sprung up in the United States in recent years. The first such station was WRNO in New Orleans, Louisiana. WRNO uses a format of rock and roll music and lets overseas listeners get a taste of what American domestic radio is like. American commercial broadcasters have generally met with modest, if any, success. (One particularly interesting attempt was KYOI, broadcasting a Japanese language rock and roll format from the Mariana Islands!) Despite this record, newcomers continue to try their luck.

Almost all international broadcasters seek contact with and letters from listeners. Sometimes this is out of a genuine desire for opinions and comments from listeners. (Several broadcasters operated by major European nations fall into this category.) Others seek listener mail to demonstrate to funding authorities that their programs are indeed being listened to in the intended target countries. More typically, listener contact is solicited so that the aims of the station's funding authorities can be furthered. As an example, a letter to a religious international broadcaster requesting a program schedule will produce a reply envelope containing several religious tracts and pamphlets in addition to the desired program schedule. Tourism information is often sent by government-sponsored broadcasters. A single letter to a government-operated broadcaster can (and often does) place the listener's name and address on the station's mailing list for several years to come. Sometimes a "hard sell" is employed; Czechoslovakia's Radio Prague once sent, during the Carter administration, a postcard opposing development of the neutron bomb to American listeners on its mailing list. The postcard was preaddressed to President Carter at the White House and required listeners only to sign it, apply a stamp, and drop it in the mail.

Why would anyone write an international broadcaster? One common reason is to request a program schedule or get on the mailing list for future ones. Other listeners may have questions about the program or country. But for many listeners, the prime motivation is to obtain a *QSL* or *verification card* from a station. "QSL" is radiotelegraph code for "I confirm," and the original intent of a QSL card was to confirm that the listener indeed heard the station. To obtain QSLs, you write the station a letter (known as a *reception report*) giving information about your reception (time, date, frequency, signal quality) along with enough details about what you heard (program titles, announcer names, music played, and the like) to convince the station you actually heard it. If your report matches the station's records for the date, time, and frequency in question, the station sends you a QSL. (Actually, the process is not this smooth nor does a QSL really "prove" someone heard a station. We'll explore the entire subject of "QSLing" later.)

FIGURE 1-3

Radio Korea's QSL card includes all the information SWLs desire—the time, date, and frequency on which the station was heard.

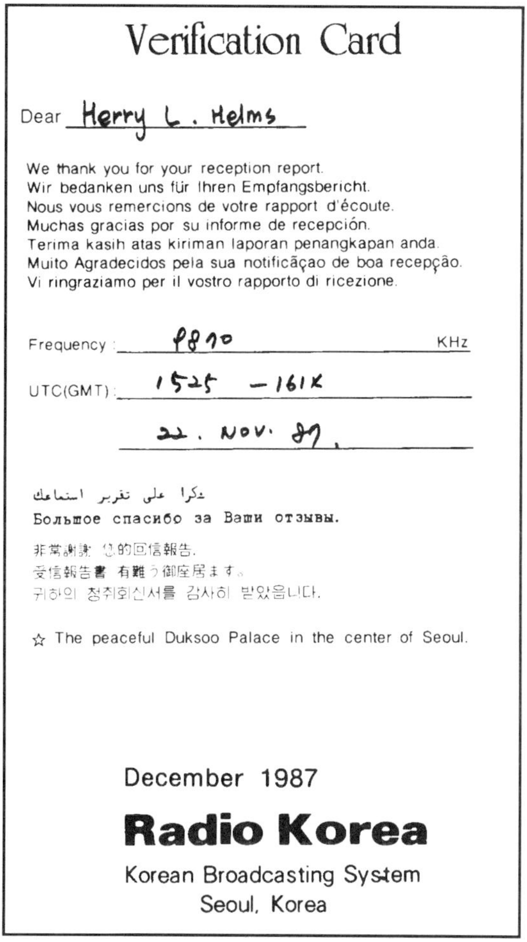

Verification Card

Dear Herry L. Helms

We thank you for your reception report.
Wir bedanken uns für Ihren Empfangsbericht.
Nous vous remercions de votre rapport d'écoute.
Muchas gracias por su informe de recepción.
Terima kasih atas kiriman laporan penangkapan anda.
Muito Agradecidos pela sua notificãçao de boa recepção.
Vi ringraziamo per il vostro rapporto di ricezione.

Frequency : 9870 KHz

UTC(GMT) : 1525 — 161X

22. NOV. 87.

شكرا على تقرير استماعك
Большое спасибо за Ваши отзывы.
非常謝謝 您的回信報告.
受信報告書 有難う御座居ます。
귀하의 청취회신서를 감사히 받았읍니다.

☆ The peaceful Duksoo Palace in the center of Seoul.

December 1987

Radio Korea

Korean Broadcasting System
Seoul, Korea

SWLs collect QSL cards and letters for the same reasons other people collect baseball cards, coins, or matchbooks—i.e., no one really knows why!

Many listeners try to collect QSLs from as many different countries and stations as possible. To keep SWLs sending in additional letters after they receive a QSL, many stations offer a series of cards that can be obtained only by sending in a number of reception reports within a specified period of time. Other QSL cards are issued for specific events, such as station anniversaries or the introduction of new broadcasting facilities. Sometimes special QSL cards are sent out for reports on new transmitter sites, when listener reports are especially valuable. Many stations take great pride in the design of their QSLs and produce colorful, artistic cards.

Other souvenirs can be collected from stations. Some stations send pennants to listeners. This practice originated among Latin American stations broadcasting to audiences within their own country, but soon spread to major international broadcasters. Some stations send pennants upon request, while others require a certain number of reports for a pennant. Some stations, such as Radio Moscow, apparently send pennants whenever the mood strikes them. In previous years, most pennants were made of cloth but more recently paper and plastic have been used. Some SWLs have managed to accumulate several hundred different pennants from various stations. Unfortunately, many stations have been forced to curtail sending pennants due to financial considerations.

Many stations use postage stamps rather than a meter on envelopes mailed to listeners. Some make it a point to use the latest issues and commemorative stamps. Some stations send out Christmas or New Year's cards to listeners and even sponsor listener "clubs," which offer certificates and "diplomas" to those who send in a specified number of reports.

For years, collecting QSLs from different countries and stations was the cornerstone of the SWL hobby. SWL clubs often featured numerical rankings of members by the number of different countries and stations they had received QSLs from. Accumulation of QSLs from a large number of different countries and stations was viewed as testimony to a SWL's skill and experience in receiving signals. Lately, however, some SWLs and personnel at international broadcasters have been critical of the entire practice of QSLing and collecting similar souvenirs from stations. Personnel at a few stations have complained that sending out QSLs and similar items is an expensive waste of money and staff which could be better utilized producing better, more interesting programs. (This is an argument which implies that creativity and imagination are commodities that can be purchased!) Some stations have even gone so far as to stop issuing QSL cards altogether, or replaced them with "listener cards" sent out to everyone who writes the station for any reason. Such listener cards look like QSLs, but make no pretense of confirming that the listener heard the station. These and related station actions have found support among some SWLs who feel that stations should be listened to for program content alone and not just to get a QSL.

Yet QSLs seem destined to remain an important part of the experience of shortwave listening. Many listeners still enjoy collecting colorful cards from stations they have heard, and such QSLs can develop considerable goodwill toward the station. Perhaps the most important argument is that stations have yet to devise a more effective means of generating listenership and listener mail than the humble QSL card. It remains a powerful positive reinforcer for desired behavior—listening to and writing the station. (Many listeners collect QSLs from stations other than international broadcasters; this will be covered later.)

Many shortwave stations are intended for reception within the country they broadcast from. Such *domestic* shortwave broadcasters are not always located in a Third World nation; countries such as Canada, Australia, West Germany, and the USSR all have domestic shortwave broadcasters. Domestic shortwave broadcasters are often more interesting than the powerful international broadcasters. You won't hear the loud, booming signals international broadcasters use, but you won't find the carefully tailored, sanitized programming either. Programs are in the local language(s) of the countries involved; in addition to Spanish, French, German, or Russian, you'll hear Swahili, Sesotho, Hausa, Pulai, and even Tahitian. You don't have to understand these languages to enjoy listening to domestic broadcasters—wait until you hear the heavy breathing on a Spanish language "radionovella" (soap opera) and the hysterical announcing style in Portuguese of a soccer match from Brazil. And music (even vocals) can be enjoyed without translation. The music you can hear is often totally unlike anything you've heard before (or could hear anywhere else). Several listeners have compiled libraries of taped music from domestic shortwave broadcasters. Domestic broadcasters give something of the texture of life in a country, since you and citizens of that nation are simultaneously listening to the same broadcast. You may be surprised at the impact—or lack of impact—of American culture upon a particular country. (I will never forget hearing Gene Autry records being played by a station in Uganda!) And you can hear where elements of American culture, particularly music, had their origins.

DXers find domestic shortwave stations to be among their favorite targets, since they are more challenging to receive. Many international broadcasters use transmitters rated at 250,000 to 500,000 watts (or, as it's usually expressed, 250 to 500 kilowatts, abbreviated KW) of power. These transmitters are connected to efficient directional antenna systems giving *effective* transmitter powers of over one million watts (one megawatt). Hearing such stations is no problem; in fact, they're difficult to avoid. By contrast, domestic broadcasters use but a fraction of the transmitter power and normally use simple antenna systems which don't boost the apparent power of the signal. It's rare to find a domestic broadcaster running even as much as 50,000 watts (50 KW) of power; 1 to 5 KW are more typical power levels. Moreover, almost all international broadcasters try to schedule their programs when reception in the intended target areas would be best. Domestic broadcasters operate according to the needs of their home populations, and the hours when they are on the air are often not the best ones for reception in North America. Thus, listeners in the United States always find it much easier to hear broadcasts from Radio Japan or Radio Beijing than Radio Mil, a domestic broadcaster in Mexico that uses only 250 watts of power.

Fortunately, many domestic stations operate in frequency ranges where few, if any, international stations are found. Three special broadcasting bands have been established solely for stations located between the Tropic of Capricorn and the Tropic of Cancer. These bands were set up because static on the AM broadcasting band is often so heavy in the tropics that reception of AM stations outside of their immediate vicinity is difficult or impossible. These so-called "tropical bands" are favorites of SWLs and DXers worldwide seeking local color or rarely heard stations. Other domestic stations can be heard better when international broadcasters are not usually transmitting to one's listening area. In North America, this is generally the period between local midnight and sunrise.

Numerous domestic broadcasters from Central and South America can be heard throughout the evening and night in North America. Several stations from Africa can also be heard

FIGURE 1-4

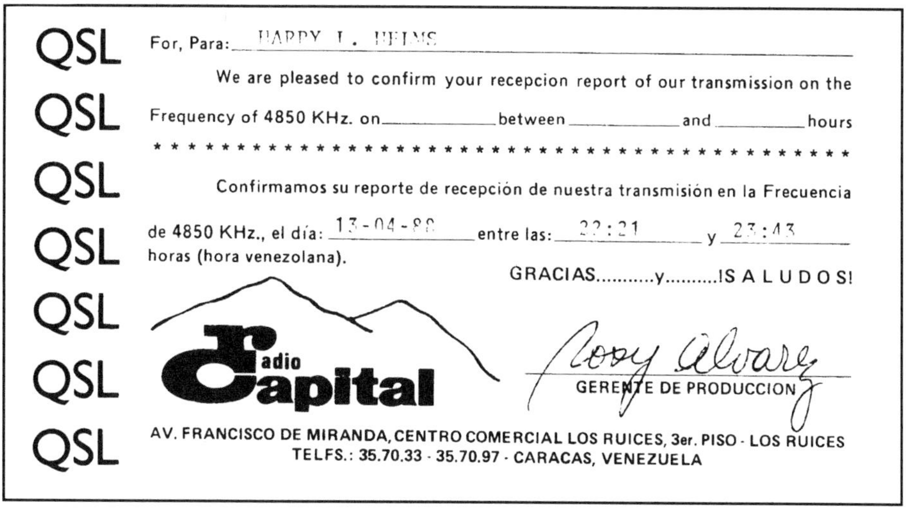

QSL QSL QSL QSL QSL QSL QSL

For, Para: HARRY L. HELMS

We are pleased to confirm your recepcion report of our transmission on the Frequency of 4850 KHz. on ________ between ________ and ________ hours

* *

Confirmamos su reporte de recepción de nuestra transmisión en la Frecuencia de 4850 KHz., el día: 13-04-88 entre las: 22:21 y 23:43 horas (hora venezolana).

GRACIAS...........y..........¡S A L U D O S!

Radio Capital

GERENTE DE PRODUCCION

AV. FRANCISCO DE MIRANDA, CENTRO COMERCIAL LOS RUICES, 3er. PISO - LOS RUICES
TELFS.: 35.70.33 - 35.70.97 - CARACAS, VENEZUELA

QSLs from domestic shortwave broadcasters, such as Venezuela's Radio Capital, are highly prized by DXers.

in North America. Perhaps the most exotic listening comes from stations located in Pacific and Asian nations such as Indonesia. Several DXers in the United States devote most of their listening time and effort toward hearing QSLing stations located in Indonesia. Brazil is another favorite target country.

Amateur "Ham" Radio

Throughout the shortwave spectrum, there are bands set aside for use by ham radio operators. Hams can be heard communicating with each other by voice, Morse code, and specialized methods such as radioteletype (messages entered at a keyboard and printed out on the receiving end) and slow-scan (still picture) television. Hams even communicate through satellites designed and built entirely by hams; these satellites have been launched by American, Soviet, and European rockets. Other hams have recently begun communicating by linking their personal computers together by radio through a system known as *packet radio*. Segments of the ham bands have been allocated to various methods of communication (voice, Morse code, and so on) by informal agreement or by law.

As mentioned earlier, hams can transmit on shortwave because they have a license issued by the government of the country in which they are located. Such licenses are issued when an applicant passes a written examination on radio theory and (usually) a Morse code receiving test. Hams are issued unique call signs to identify their stations when they're on the air, and become associated with and referred to by their call signs. Ham call signs usually consist of one or two letters, a digit, and one to three additional letters. The letter or letters preceding the digit indicate which nation licensed the station; the alphabet is divided up for this purpose by international agreement. Call signs beginning with W to WZ, K to KZ, N to NZ, or AA through AL always belong to a station licensed by the United States. Similarly, a call sign beginning with G to GZ belongs to a station licensed by Great Britain. A list of these call sign allocations is included in the appendix of this book. (By the way, other types of stations use this system; that's why broadcasting stations in the United States have call letters beginning with W or K.)

Conversations on the ham bands tend to be among the more interesting or more inane you'll ever hear; this is known as "ragchewing" among hams. Many hams also like to contact as many different countries as possible, and swap QSL cards with those stations to prove the contact took place (this is also known as DXing). Other hams try to contact all U.S. states or counties. On-the-air contests are popular, with amateurs trying to contact as many different countries, "zones," or other geographical and political divisions as possible.

Much of the activity on the ham bands might seem pointless, but the situation changes during emergencies. During 1989, ham operators were the only functioning communication links in the first hours after the San Francisco earthquake and Hurricane Hugo disasters. When trouble hits, the nonsense stops and hams provide vital communications services that civil and military authorities are unable to provide. During such emergencies, tuning the ham bands can let you hear information directly from the source and ahead of established news media.

Some SWLs get so fascinated at listening to hams that they eventually obtain their own amateur license. A few SWLs specialize entirely in listening to hams; they send reports to hams and collect QSLs from them just as avidly as other SWLs do with broadcast stations.

Utility Stations

Most stations you hear on shortwave are not broadcasters or hams. They are *utility* stations. As the name implies, these stations do some type of work, such as communications with ships at sea or aircraft aloft. Utility stations also include the communications networks of military forces, relay stations, and navigation beacons. They are not generally intended for reception by the public, although they are often heard with little trouble.

Like hams, utility stations use voice and Morse code transmission. Many utility stations use radioteletype. Another method is *facsimile*, similar to the "fax" machines in offices (on shortwave, it's mainly used to send weather maps and charts to ships at sea). Other utility stations use esoteric methods, such as *multiplex*, to send multiple signals on a single frequency. Unfortunately, intercepting and decoding these latter signals are beyond the technical and financial capabilities of most SWLs.

There are four major types of utility stations. *Fixed* stations operate from a specific land location and are used mainly to communicate with other fixed stations. *Land mobile* stations also operate from land, but from different locations or while in motion from place to place. *Maritime* stations operate from ships or are land stations used exclusively to communicate with ships. *Aeronautical* stations operate from aircraft or transmit to planes in flight.

Some utility stations are operated commercially by such companies as Great Britain's Cable & Wireless, Ltd., and derive revenue by charging for the messages they handle, much like a telephone company. A few companies even operate utility station networks devoted solely to handling messages between units of the company. But the vast majority of utility stations are owned and operated by various governments, with these stations falling into the broad classifications of "civilian" and "military."

Government utility stations are often used to facilitate transportation of some sort, such as sea or air travel. You can listen to aviation weather broadcasts from airports and aircraft aloft along with transmissions to and from aircraft flying international routes. Ships at sea use shortwave radio to communicate with seaports and other ships. Many United States government agencies use shortwave radio to provide communications during emergencies and to back up existing telephone and telex systems. Another significant use of shortwave by the U.S. government is for law enforcement; the Federal Bureau of Investigation, Drug Enforcement Administration, and Customs Service all maintain utility stations that SWLs can eavesdrop on.

Even communications associated with the President and Vice President of the United States can be heard. Many SWLs have been able to hear phone calls placed by the President while aboard Air Force One. Most sensitive communications are scrambled, but sometimes interesting conversations are transmitted "in the clear." An example took place in 1985, when U.S. jets intercepted an Egyptian airliner transporting terrorists and diverted the plane to Sicily. Several SWLs managed to hear "Rawhide" (the code name used by President Reagan) discuss plans for the operation with Secretary of Defense Caspar Weinberger as Air Force One flew back to Washington. Such security lapses are rare but keep a small but enthusiastic group of listeners glued to the frequencies used by Air Force One.

International organizations maintain utility stations and networks. The International Red Cross is one, as is the International Police Organization (Interpol). SWLs can listen in on their activities. Diplomatic services of various nations make use of shortwave facilities, nominally to keep in touch with their home governments.

A useful type of utility station is the *standard time and frequency* station. These stations are operated by various governments on precisely maintained radio frequencies; they also transmit highly accurate time signals (usually obtained from atomic clocks). In the United States, the National Bureau of Standards operates two such stations, WWV in Colorado and WWVH in Hawaii. Both stations can be easily heard through-

FIGURE 1-5

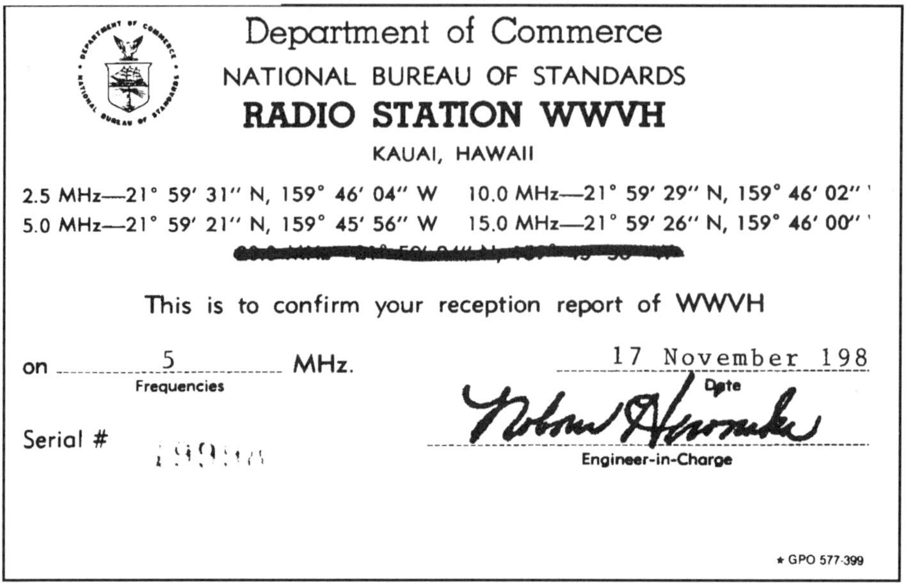

Department of Commerce
NATIONAL BUREAU OF STANDARDS
RADIO STATION WWVH
KAUAI, HAWAII

2.5 MHz—21° 59′ 31″ N, 159° 46′ 04″ W 10.0 MHz—21° 59′ 29″ N, 159° 46′ 02″
5.0 MHz—21° 59′ 21″ N, 159° 45′ 56″ W 15.0 MHz—21° 59′ 26″ N, 159° 46′ 00″

This is to confirm your reception report of WWVH

on 5 MHz. 17 November 198
Frequencies Date

Serial # [illegible]
Engineer-in-Charge

★ GPO 577-399

Standard time and frequency stations, such as Hawaii's WWVH, also issue QSL cards for correct reception reports.

out North America with their distinctive voice announcements of the time each minute. Other such stations are scattered throughout the world.

Military forces of all countries make extensive use of shortwave. Like their civilian counterparts, armed forces use shortwave to keep in touch with aircraft and ships. In addition, shortwave is also used for communication among separated land forces. The various branches of the United States military may comprise the largest single block of shortwave stations in the world. American military stations range from Coast Guard outlets rendering assistance to ships in distress to coded messages transmitted to Strategic Air Command bombers in flight. (Surprisingly, many U.S. military stations will send QSLs in response to reception reports.) The military forces of other countries can also be heard on shortwave, and it is not uncommon for listeners in North America to hear the military forces of Latin American nations (including Cuba). Listeners in Europe often run across signals from Soviet military forces. Identifying which (or what) military station you're hearing can be a challenge, since so-called "tactical" call signs (such as "Thunderchief") are often used along with sophisticated speech-scrambling and encoding methods. But many listeners find such problems part of the fun of tuning military and other utility stations.

AM Broadcast Band DXing

As I mentioned in the preface, my interest in radio was triggered by the fact that I could hear stations on the standard AM broadcasting band from over one thousand miles away at night but not in the daytime. You can observe this for yourself. Tune across the AM dial at your local noon. Do the same at midnight. You'll hear more stations at midnight, with many of them badly interfering with each other, and some of those stations will be from hundreds or even thousands of miles away.

Several listeners devote all their time to DXing the AM broadcast band. Some of them have managed impressive results, such as hearing over 100 different countries on the AM band. With proper receiving conditions, equipment, and listening skill, they have accomplished remarkable feats of reception. Australia has been heard on the East Coast of the United States on AM, for example, and Europe has been heard on the West Coast and in Hawaii. AM DXers on the East Coast routinely hear stations in Europe and Africa, while West Coast listeners often catch stations in Asia and the Pacific. Both coasts have opportunities to hear Central and South America.

Other AM DXers specialize in receiving stations from the United States and Canada. Their goal is to hear (and usually get a QSL from) at least one station in each state and province; most also try to hear and QSL as many different stations as they can.

FIGURE 1-6

January 8, 1988

Harry L. Helms

Dear Friend,

Thank you for your recent reception report. We are enclosing a QSL card for your collection.

WHAS is a clear channel station. This means that at night we are the only station in the United States transmitting on 840 KHz. Our power is 50,000 watts both day and night, and our antenna is a single 664-foot tower. WHAS began operation in 1922, making it one of the first broadcast stations in the world. We are the only station in Louisville broadcasting in AM Stereo.

Reception reports have come to us from twenty-eight countries on six continents, as well as from every state in the USA and every province in Canada. Outside North America, the largest number of reports have come from Finland and Sweden.

We are always pleased to get reports from our listeners. We hope you have enjoyed listening to WHAS, and that you will be able to listen again in the future.

Best wishes,

Charles R. Strickland
Chief Engineer

520 W. CHESTNUT STREET • PO BOX 1084 • LOUISVILLE, KENTUCKY 40201 • (502) 582-7840

WHAS on 840 kHz is one AM broadcast band station that's obviously proud of the reports it's received from listeners around the world!

As a general rule, it's more difficult to hear an AM broadcast station than a shortwave station over the same distance. Reception conditions (which we'll discuss later) play a major role in AM DX reception; conditions necessary for outstanding reception (such as of Australia on the East Coast) may be present only a few days per year (or may not be present at all during years when the sun is especially active). Moreover, better receiving equipment and antennas are necessary for AM reception over distances comparable to shortwave, and the level of interference is usually much greater. This is not to imply that AM DXing is superior to SW DXing; it just means they are different. (SW DXing is often just as demanding, and it is possible to hear stations on shortwave frequencies at distances and power levels which are impossible on AM.)

But you don't need exceptional conditions or equipment to get started in AM DXing. Any radio you happen to have—portable, stereo receiver, clock radio, and the like—is capable of pulling in distant stations at night. Several of the skills developed in AM DXing are useful in SWLing, making the AM band a good (and inexpensive) place to start in the hobby of SWLing.

FM Broadcast and TV DXing

When channels were allocated for television broadcasting and FM radio, the frequencies were assigned in the belief that they were relatively free of the reception conditions that make possible long distance reception on the AM and shortwave bands. Interference from other stations could significantly degrade picture and sound quality, and the FCC went to considerable lengths (including carefully spacing apart stations using the same channel or frequency) to minimize the possibility of interference.

By and large, the FCC did its job well. You can see this for yourself by tuning across the AM broadcast band, FM broadcast band, and the TV channels at midnight your local time. You'll *normally* find the same stations on FM and TV channels you can hear at noon; reception will be clear with little interference (if any) from stations outside your local area. In contrast, the

AM band will be a cacophony of distant and local stations crashing against each other.

The key word in the last paragraph was "normally." On several occasions each year, freak atmospheric conditions make it possible to receive FM and TV stations located hundreds or even thousands of miles away. If you have a TV station in your area on channels 2, 3, or 4, you may have seen rolling black bars across the screen during the months of June and July. (The station may even make an announcement that your TV set is not at fault.) Such rolling black bars are caused by distant stations on the same channel trying to break through, and indicate that conditions are right for reception of distant FM and TV stations. FM and TV DXers refer to this as a "band opening," or simply as an *opening*.

The conditions which permit FM and TV DXing are unpredictable, although they are more likely to occur during certain times of the year and during years of high solar activity. This means that you can't plan for FM and TV DX reception the way you can for shortwave; you have to be lucky enough to be listening or viewing when conditions are right. Even when conditions permit DX on FM and TV, the conditions may (and often do) change rapidly and in unexpected ways. For example, some conditions for distant receptions may last for a week, while other conditions may last only a few minutes or less. Stations can show up unexpectedly for just a few seconds of reception before disappearing. (This happens when FM and TV signals are reflected off the ionized trails left by meteors entering the Earth's atmosphere.) And the *direction* from which distant TV and FM signals can be heard often changes during an opening. For example, distant TV and FM stations may be heard at first from the west of your listening location. As the opening progresses, stations to the west may abruptly vanish and be replaced by stations to the south of your location. After a few minutes, all distant stations may be gone and the FM band and TV channels back to normal.

Such unpredictability makes FM and TV DXing frustrating to many. However, it means success depends more upon the ability to recognize unusual reception conditions than it does on equipment. Any ordinary FM radio or TV set can be used. A sophisticated antenna system isn't necessary; often a pair of TV "rabbit ears" or the built-in telescoping antenna on an FM portable outperforms a large TV/FM antenna mounted on a rooftop. FM and TV stations also send out QSLs for reception reports from distant listeners and viewers. Unlike shortwave broadcasters and larger AM stations, FM and TV stations are not flooded with reports and often are genuinely pleased to learn they have been heard or seen many miles away.

Clandestine, Pirate, and Illegal Radio

Not all stations you hear on shortwave operate under international agreements or national laws. Scattered throughout the shortwave spectrum are various stations which operate from hidden locations or in violation of national or international laws. If such stations are extralegal broadcasters with political overtones or purposes, they are known as *clandestine* stations. If they are non-political or "hobby" broadcasters, they are called *pirates*. The rest, ranging from unlicensed ham-type operations to drug smuggling networks, fall into the broad category of *illegal* stations.

Clandestine broadcasters are almost always operated by a government or with the support of a government. The aim is to influence (and sometimes incite) the population of a target country. An element of deception is normally present in clandestine operations. For example, the clandestine might pretend to be actually operating from within the target country. The true sponsor and purpose of the clandestine station are usually concealed, and a fabricated "cover" story may be used instead. The content of clandestine broadcasts is highly political, although the politics may be blended with music and other

FIGURE 1-7

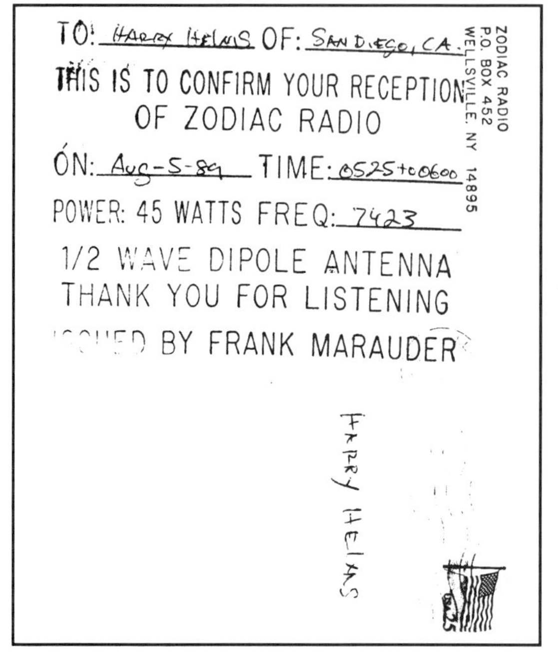

A QSL card from a station that was! Zodiac Radio was an illegal broadcasting station operating from Anaheim, California. Unfortunately for the operator, the Federal Communications Commission located and closed his station in early 1990. The operator also received a hefty fine from the FCC!

features to attract listeners. Clandestines appear and vanish according to shifting political currents. For example, Nicaragua was the target of much clandestine activity during the final days of the Somoza regime. (Listeners to one such clandestine, Radio Sandino, were instructed on how to fire an automatic rifle and make Molotov cocktails.) After the Sandinistas assumed power, new clandestines opposed to them sprung up.

Pirate broadcasters are privately operated, often as a hobby, in violation of the laws of the country in which they are located. They are generally low-powered and noncommercial; they operate at irregular hours, and the operators are usually young. In the United States and Western Europe, pirate broadcasters typically play much rock music, run satirical skits (some of which are genuinely hilarious), and get obscene at times. The content may be political, but seldom is seriously so. (Usually politics are limited to calls for "free radio," legalizing marijuana, and similar pressing concerns.) The atmosphere is often one of a group of young people having a party that just happens to be on the air for all to hear.

One fascinating aspect of American and Canadian pirate stations is that several have been operated by SWLs apparently tired of just listening to stations! Some SWLs have received telephone calls from pirate operators announcing they were about to go on the air. Other SWLs and DXers have reported their receptions of various pirate stations to the bulletins published by SWL clubs, and have unexpectedly received QSL cards from the pirate operators who were apparently members of the same clubs.

Pirate broadcasters are also found in other nations. In the Soviet Union, such stations as known as "radio hooligans." They broadcast Western rock music and irreverent jokes about the system. In some Third World countries, such as Indonesia and Thailand, some pirates operate almost as a public service for small, isolated communities.

Illegal stations are engaged in two-way communications rather than broadcasting. Converted or modified ham radio equipment is normally used. One recent growth area for illegal radio has been drug smuggling networks. SWLs (and various law enforcement agencies) have listened in on radio networks coordinating drop shipments from ships or planes to delivery points on land. Radio is also used for communications between drug growing or processing areas and places where shipments are dispatched.

Guerilla groups in Latin America make extensive use of two-way radio. In fact, a major magazine article on Eden Pastora, a leading anti-Sandinista military commander, showed him seated in front of a modern ham radio "transceiver" (a combination transmitter/receiver). During the final months of the Sandinista campaign against the Somoza government, many SWLs in North America heard cryptic communications in Spanish that were of a military nature. Paramilitary organizations in the United States are also reported to use two-way radio for their activities.

Not all illegal radio is so sinister. Many are stations talking to each other in a manner similar to ham radio operators—but without a license and on frequencies reserved for other stations, such as utilities. Many fishing boats can be heard on illegal radio networks talking about the day's catch and related topics such as sex and drinking.

Listeners can also hear mysterious "numbers" stations. These do little more than transmit groups of numbers, usually using a woman's voice, in languages such as Spanish, English, and German. It's believed these transmissions are some form of coded messages from intelligence agencies to their operatives in the field. The entire subject of "numbers" stations will be discussed later.

The Shortwave Listening Hobby

Many shortwave listeners like to keep in touch with other persons who share their interest in SWLing. These people do more than casually listen to a handful of favorite SW stations; instead, they want to keep up-to-date on schedules and frequencies used by stations and improve their receiving capabilities. Their approach to SWLing is serious enough that it can legitimately be classified as a "hobby," much like an "audiophile" who takes audio equipment and technology more seriously than the typical person with a home stereo system.

Since the early days of radio, several clubs and organizations have been formed to allow SWLs to make contact with other listeners, to swap news about what is being heard, and to serve as a forum for opinions about receiving equipment, favorite or least favorite stations, and related topics. Most of these clubs, all operated on a nonprofit basis by unpaid volunteers, have banded together under an umbrella organization known as the Association of North American Radio Clubs (ANARC). ANARC holds an annual convention for SW enthusiasts, and some of the ANARC member clubs hold their own annual conventions. Clubs run the gamut from the so-called "all band" clubs (those covering shortwave broadcast, utilities, AM/FM/TV DXing, and so on) to those that specialize in a specific area such as shortwave broadcast, AM DXing, and even pirate and clandestine radio.

These clubs serve as the focal point for the shortwave hobby. Although there is no requirement to join a SWL club, most serious SWLs eventually do join one or more clubs according to their interests. Clubs publish bulletins on a monthly (or more frequent) basis, and these bulletins are the best source of current news relating to shortwave activity. Club bulletins contain information on which stations are being heard, changes in

FIGURE 1-8

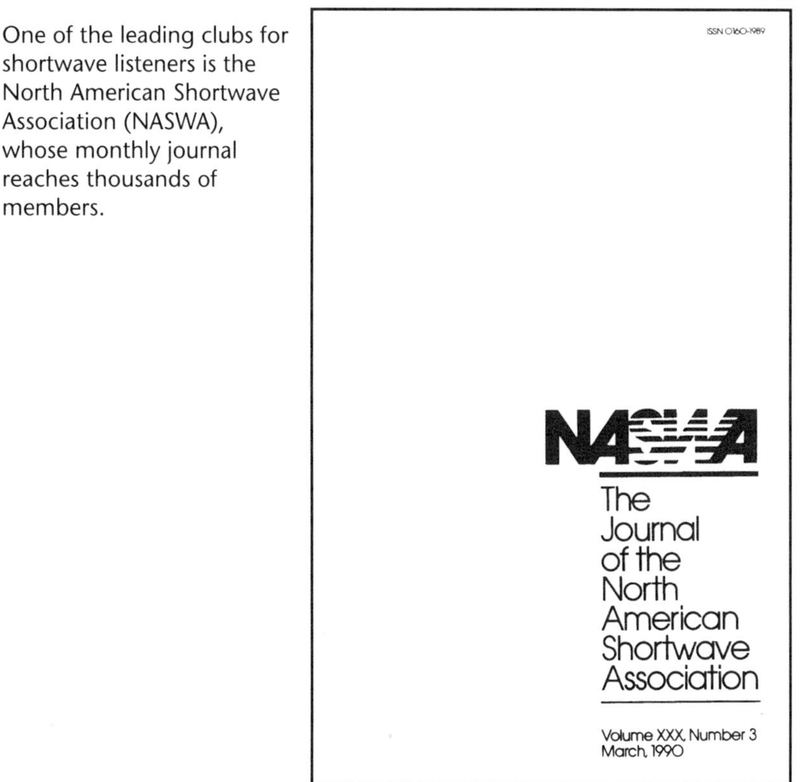

One of the leading clubs for shortwave listeners is the North American Shortwave Association (NASWA), whose monthly journal reaches thousands of members.

frequencies and times used by broadcasters, and details of the latest QSLs received by members. Bulletins often contain member evaluations of new shortwave receivers and accessories along with articles by more experienced members offering tips on improving shortwave reception, identifying stations broadcasting in foreign languages, and similar subjects.

Local and regional SWL groups have also been formed. These are largely social in nature, and provide a forum for listeners to swap listening experiences, to view QSL collections, and to discuss listening techniques. Some SWLs also form telephone alert groups, in which any listener who notes a band opening or a rare station calls other members and informs them of the news.

A recent innovation has been the use of computer "bulletin boards" and "teleconferencing" systems to link together SWLs. These have grown rapidly in popularity as they are an effective means to spread news widely, and many SWLs enjoy the opportunity to "listen together" with other SWLs scattered across the country using personal computers.

Other SWLs rely on commercial publications for information. The most notable of these is *Popular Communications*, which is available at many newsstands. Another is *Monitoring Times*, available at many electronics and ham radio equipment dealers. There are also several newsletters devoted to selected topics in SWLing (such as "numbers" stations). A major difference between clubs and such publications is that clubs usually have more recent news and information, since the press deadline for club bulletins is only a few days instead of several weeks. Clubs consisting of a few hundred members can provide greater contact among listeners than a publication with many thousands of readers. Yet commercial publications are preferred by some SWLs, since they can run photographs and illustrations which club bulletins often avoid because of cost. The editorial quality of commercial publications is also higher, since they are edited and managed by publishing professionals. Finally, commercial publications are generally free of the personality conflicts and club politics which can take up much space in some club bulletins. For many listeners, a combination of club memberships and a subscription to one or more commercial publications allows them to keep up to date on the latest news relating to their interests.

Most listeners are content to casually pursue the hobby, but others decide to take SWLing very seriously. Some try to QSL literally every station they hear; if a station doesn't reply after their first reception report, they keep sending additional reports until they get a reply—even if it takes over a dozen reports. Others try to collect as many different QSL card designs as possible, sending the same group of stations a report each month. Specialization is common, with some SWLs devoting all their time and energy to shortwave broadcast, utility, or FM station reception. Perhaps the most serious are those DXers trying to hear and QSL as many different countries as possible. They almost always have invested considerable money in their receiving equipment and think nothing of crawling out of bed at 3:30 a.m. on a workday so they can try to hear a 500-watt station located in Indonesia or some other exotic locale.

The degree of hobby participation is up to you. Many, perhaps most, SWLs never join any club or read a commercial magazine; they enjoy tuning the bands on their own and discovering new stations and frequencies. A few become so wrapped up in the hobby that it becomes the major focus of their lives (even to the extent that almost all of their friends are SWLs). Again, SWLing is what you want it to be and what you make it.

Terminology, Timekeeping, and Related Matters

Like other specialized interests, SWLing has developed its own terminology and nomenclature. You'll find these terms and expressions used in club bulletins, publications, and by radio stations in their correspondence. We've already introduced some of them, and there are others you'll often see. Like "QSL," many originated as radiotelegraph codes and are still used by hams for that purpose. One is "QRM," which stands for *interference*. Another is "QRN," which means *noise* from thunderstorms, power transformers, neon lights, and the like. The appendix of these book contains a list of common "Q-signals." Finally, you'll often hear shortwave broadcasters and hams using the term "73" on the air and in correspondence. "73" is a traditional radiotelegraph code for "best regards," and is used among stations, SWLs, and hams as a friendly way to end a broadcast, contact, or letter.

Since shortwave radio crosses international boundaries, a worldwide standard of timekeeping is necessary. For this, *coordinated universal time* (abbreviated UTC from the French term for it) is used. You might be familiar with this system under its old name of *Greenwich mean time* (GMT). This method uses the 0° meridian at Greenwich, England as the standard reference for defining a time for use throughout the world. This means you'll have to add or subtract hours to or from UTC, depending on where you're located in the world, to determine the local equivalent time to UTC. The advantage is that a time and day such as "2200 Wednesday UTC" means the same whether you're listening from Tokyo, Paris, or Chicago. All international broadcasters use UTC in their program schedules and generally want time in reception reports to be indicated in UTC. UTC is also the standard time used by hams and utility stations. Moreover, times in SWL club bulletins and commercial publications are almost always UTC (the exceptions tend to be clubs specializing in AM, FM, or TV DX).

If you're listening from North America, you can convert UTC to local time by *subtracting* the proper number of hours from UTC. The conversions in Table 1-1 are for major North American time zones:

UTC uses the 24-hour or "military" system of time notation. In UTC, midnight is given as 0000. The next hour (or 1:00 a.m. at the Greenwich meridian) is written as 0100. The time

TABLE 1-1

Time Zone	Subtract from UTC for Local Time
Atlantic Standard	Four hours
Atlantic Daylight	Three hours
Eastern Standard	Five hours
Eastern Daylight	Four hours
Central Standard	Six hours
Central Daylight	Five hours
Mountain Standard	Seven hours
Mountain Daylight	Six hours
Pacific Standard	Eight hours
Pacific Daylight	Seven hours
Alaskan Standard	Nine hours
Alaskan Daylight	Eight hours
Hawaiian Standard	Ten hours

fifteen minutes later is expressed as 0115. This system continues with 0200, 0300, 0400, and so on until UTC afternoon is reached. The next minute following 1259 UTC (or 12:59 p.m. at Greenwich) is 1300 UTC. The time continues with 1400, 1500, 1600, and so on until 2359 UTC is reached; one minute later is 0000 UTC and the start of a new UTC day.

An important point to remember is that UTC refers to the *day* as well as the time. For example, if you want to hear a broadcast scheduled for 0300 Wednesday, and you live in the Eastern Standard time zone, you would listen at 10:00 p.m. *Tuesday*. Forgetting to make the necessary day conversion in addition to the time conversion is a common error when using UTC.

Throughout this chapter, we've used the terms "radio" and "receiver." What's the difference between a "shortwave radio" or "shortwave receiver"? Nothing really, although a receiver is usually thought of as a more complex and versatile device than a radio.

CHAPTER TWO

Understanding the Shortwave Bands

A LOT OF PEOPLE PANIC the first time they tune a shortwave radio. Instead of the order found on the AM and FM radio bands, they hear a chaotic jumble of strange signals and noises without any apparent organization. But this is just an illusion. The shortwave spectrum has been carefully divided up into bands allocated for specific purposes. In this chapter, we'll look at these bands and some of the things you can hear on them. We also look at the different methods of transmission, sometimes known by the technical name *modes of emission*, used by shortwave stations.

Two terms familiar to every SWL are kHz *(kilohertz)* are MHz *(megahertz)*. These are the two units in which radio frequencies are measured. (These terms replace two older units you may have heard of called "kilocycles" and "megacycles.") A *Hertz* (Hz) is one cycle of a radio wave (a cycle is the peak of a radio wave through to the peak of the next wave) One kilohertz is equal to 1000 Hertz, and one megahertz equals 1000 kilohertz. The conversion between these units is:

5 MHz = 5000 kHz = 5,000,000 Hz

As a practical matter, you only have to be concerned with kHz and MHz. With a little practice, you'll soon be able to convert between 4895 kHz and 4.895 MHz without trouble. Both kHz and MHz are used for station frequencies, although kHz is commonly used for frequencies from 0 to 30,000 kHz

(30 MHz) and MHz is used for frequencies above 30 MHz. That convention will be followed in this chapter and throughout the rest of this book.

Modes of Emission

You're already familiar with the terms "AM" and "FM." These are two emission modes used to transmit voice and music. There are several others used on shortwave. This section will discuss them in a descriptive, simplified fashion. If you're technically inclined and want more details, consult a book such as *The Radio Amateur's Handbook* published by The American Radio Relay League.

All radio signals take up some frequency space, known as *bandwidth*. The bandwidth occupied by a signal varies with the mode. As a general rule, the more complex and "information packed" a signal is, the greater the bandwidth it occupies. For example, a Morse code signal is very simple; it is generated simply by turning a radio transmitter off and on to form the Morse characters. Since it is so simple, a Morse signal usually occupies a bandwidth of a few hundred Hz. On the other hand, a stereo signal from your local FM broadcast station can occupy over 60 kHz of frequency space. This is because the FM station broadcasts music over a wide frequency range in stereo, and there's much more "information" in such signals than the dots and dashes of the Morse code.

If you look through the frequency listings of *Passport to World Band Radio* or a similar publication, you'll note that most shortwave stations operate on frequencies spaced five kHz apart from each other, such as 4800, 4805, 4810, 4815, 4820 kHz, and so on. This is because AM *(amplitude modulation)* stations occupy at least three kHz, and often more, of bandwidth on both sides of their listed frequency. For example, an AM station operating on 9800 kHz may actually occupy the bandwidth beginning at 9797 kHz and running through 9803 kHz. The listed frequency of 9800 kHz is known as the *center frequency* or *carrier frequency*.

If you were to tune for this station on your shortwave radio, you would set your receiver's dial to 9800 kHz for best reception. If you tuned slightly off the center frequency—to 9798 or 9802 kHz—you could still hear the station, but the audio would probably be distorted and the signal wouldn't be as strong.

Morse code sent by turning a transmitter on and off to form the different characters is known as *continuous wave* (CW) telegraphy. As shown in figure 2-1, a CW signal on a center frequency of 5000 kHz occupies very little bandwidth, often less than 100 Hz on either side of 5000 Hz.

CW has several advantages in addition to its narrow bandwidth. CW transmitters are simpler and less expensive than those needed for voice or other more complex modes. CW is also the most efficient mode. This is partly because of the narrow bandwidth; 100 watts of transmitter power concentrated in a 200 Hz bandwidth is more effective than the same amount of power spread over several kHz. Moreover, all transmitter power in CW goes into producing signal "information." Transmitter power in some other modes goes into producing signal components that don't carry useful information.

When a CW transmitter is left on, the signal it transmits is sometimes referred to as a *carrier*. The power output level, or amplitude, of this carrier is essentially constant when Morse

FIGURE 2-1

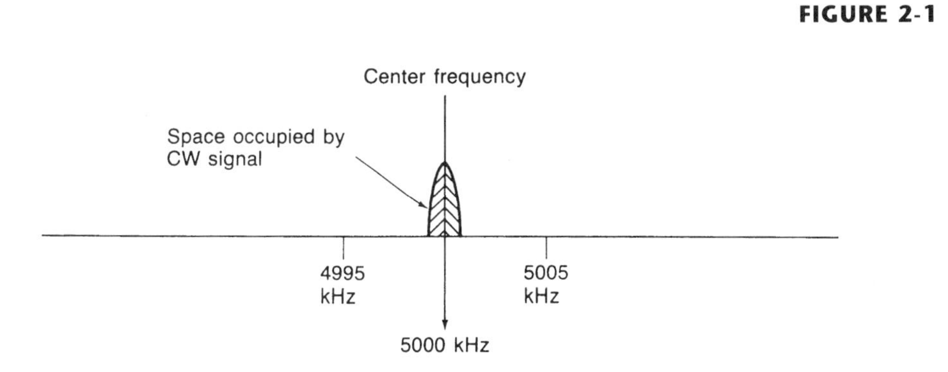

A narrow CW signal occupies only a few hundred Hz of frequency space.

code is sent. However, it is possible to make the output level of a transmitter rise or fall according to the sound patterns of voice and music. This is known as *amplitude modulation* (AM). The amplitude of an AM signal ranges from zero to approximately twice the value of the unmodulated carrier, and the amplitude changes continuously as the voice or music changes.

Figure 2-2 shows an AM signal with a center frequency of 5000 kHz. Note that the AM signal has a set of *sidebands* above and below the carrier frequency. These sidebands contain the information (voice, music, and so on) present in the AM signal, and are each equal to the highest audio frequency used to vary (modulate) the CW signal. In figure 2-2, an audio signal of 3000 Hz is used for modulation; the sideband found from 4997 to 5000 kHz is the *lower sideband* while the sideband from 5000 to 5003 kHz is the *upper sideband.* At the center frequency is a carrier. Both sidebands contain identical information (the 3000 Hz audio signal) and can be thought of as mirror images of each other. All the useful information in an AM signal is contained in either sideband; the carrier contains no information itself. The transmission of the two sidebands is a by-product of the

FIGURE 2-2

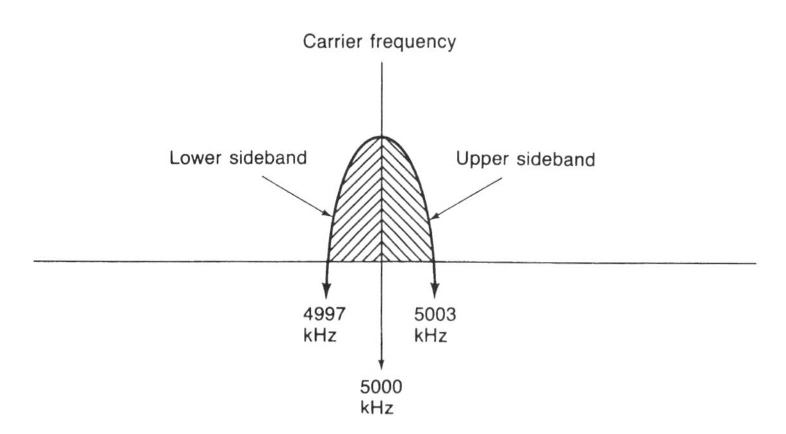

An AM signal has "mirror image" sidebands above and below the carrier frequency.

amplitude modulation process, not a deliberately-sought result. The bandwidth of an AM signal is equal to approximately twice the highest audio frequency used; if an audio tone of 5000 Hz were to be transmitted, the bandwidth would be roughly 10 kHz.

AM is easy to produce at the transmitter and simple for a listener to tune. Unfortunately, it's an extremely wasteful mode in terms of transmitter power and bandwidth. Approximately two-thirds of an AM transmitter's power goes into the carrier, which contributes no information. Only one sideband is needed, but two are produced. Of all the components of an AM signal, all that is really needed to convey the total signal information is either sideband.

This is the rationale behind *single sideband* (SSB) transmission. In SSB, a weak AM signal is generated and the carrier and one sideband are removed. The remaining sideband, either the *upper sideband* (USB) or the *lower sideband* (LSB), is amplified and transmitted by itself. The result is a voice or music signal that far more efficiently uses the available transmitter power than an AM signal; a typical SSB signal has the same efficiency as an AM signal of three to four times the transmitter power. An SSB signal occupies approximately half the frequency space of an AM signal since the other sideband is not used, and thus two SSB signals can fit into the same frequency space (bandwidth) used by one AM signal.

Since SSB sounds terrific, you might wonder why almost all broadcast stations still use AM. The answer lies on the receiving end of a SSB signal. While the carrier in an AM signal doesn't contain any information, it plays an important role when receiving (or "detecting") the signal. In a receiver, the sidebands "beat" against the carrier to produce intelligible voice and speech. Tuning an SSB signal with an AM receiver produces a distorted, unintelligible sound that some listeners find similar to Donald Duck's voice.

The solution is to produce a replacement carrier within the

receiver itself. This has traditionally been done with a circuit known as a *beat frequency oscillator* (BFO). A BFO is actually an extremely low-power transmitting circuit in the receiver producing a CW signal. The received SSB signal is then "beat against" the BFO signal so the receiver can then detect the signal in the same manner as an AM signal. The BFO is also used in receiving Morse code sent via CW; without the BFO, all you hear with a CW signal is a "thumping" sound as the background noise is quieted by the dots and dashes of the code.

Unfortunately, for many years BFOs were crude circuits except on very expensive receivers. BFOs on most receivers were prone to drift in frequency, resulting in a signal that was readable only part of the time. Moreover, other receiver qualities are more exacting for SSB reception than AM. Happily for SWLs, receiver technology has advanced to the point where almost all contemporary receivers are capable of excellent SSB reception. The availability of satisfactory SSB receiving equipment at a reasonable cost has resulted in a few international broadcasters, such as Radio Sweden, beginning some transmissions in SSB. Other broadcasters can be expected to join them in the years ahead. However, AM will still be the most widely used mode for SW broadcasting for the foreseeable future.

Outside the broadcasting bands, SSB rules for voice communications. Utility stations and ham operators have been using it almost exclusively for years.

In both AM and SSB transmission, information is transmitted by varying the output of the transmitter in accordance with voice or music. While the transmitter output changes, the carrier frequency remains constant. It is also possible to transmit information by keeping the transmitter's output constant but varying the transmitter frequency in accordance with voice or music. This is known as *frequency modulation* (FM). When no audio is being transmitted, the frequency of an FM signal rests at a *center frequency*. As shown in figure 2-3, the frequency of an FM signal varies above and below the center frequency as the

FIGURE 2-3

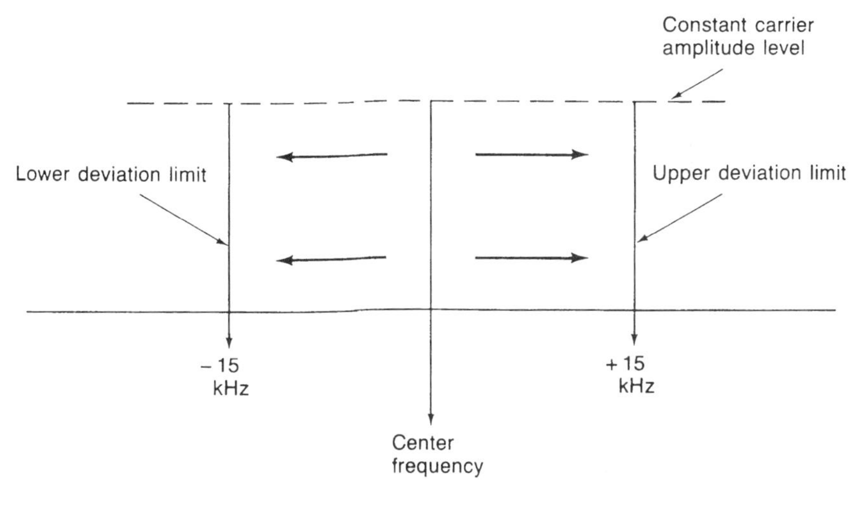

The amplitude of an FM signal remains constant. Information is transmitted by "swinging" the carrier above and below the center ("unmodulated") frequency.

voice or music being transmitted changes. The maximum amount by which an FM signal varies above or below the center frequency is known as the *deviation* of the signal. The minimum deviation of an FM signal is usually 5 kHz; since deviation is measured above *and* below the center frequency, the bandwidth of the signal would be 10 kHz. Other common deviations for FM signals include 15 and 30 kHz, meaning such signals have a bandwidth of 30 or 60 kHz. Obviously, many more AM and SSB signals can fit into the frequency space required for even a single FM signal.

Since FM needs so much frequency space, why is it used? The main reason is that radio and electrical noise (QRN) is primarily amplitude modulated. An FM receiver "ignores" the AM noise pulses, meaning FM communications have much less noise than AM. (Ever notice how "quiet" the FM broadcasting band is compared to the AM one?) Also, FM receivers exhibit what is known as the *capture effect*. When two or more FM signals are present on the same frequency, an FM receiver tends to respond only to the strongest signal and ignore all others. (This effect is more pronounced if one signal is significantly stronger than the others.) The combination of the capture effect and the general immunity to noise means that FM is capable of much better audio quality, particularly for music, than AM or SSB.

Radioteletype (RTTY) is a mode in which the transmitting station sends messages at a keyboard similar to that on a typewriter or computer terminal. These messages are transmitted by *frequency-shift keying* (FSK), a method that is something of a cross between CW and FM. At the receiving end, FSK signals are converted back to written form and are displayed on a video screen or printed out on a printer similar to that used with personal computers. Some RTTY systems combine the keyboard and printer into a single unit, and there are interface devices and software available which convert personal computers into RTTY systems.

When a message is entered at the keyboard of a RTTY transmitting station, the various letters, numerals, and punctuation symbols are converted into a code for transmission. The oldest and still most widely used is *Baudot*. Recently, the *ASCII* (American Standard Code for Information Interchange) code has become popular; ASCII is used for data exchange between personal computers as well. Both Baudot and ASCII are binary codes, meaning that all characters in each can be represented by combinations of two different conditions (much as all of the Morse code is represented by using only dots and dashes). When using FSK, the transmitter output remains constant but switches between two separate radio frequencies that represent the two different conditions. The spacing between the two frequencies is known as the *shift* of a RTTY signal; common values include 170 and 850 Hz.

It is also possible to transmit RTTY using AM or FM. Two audio tones are used to represent the two different conditions, producing a "tweedling" sound as the signal alternates between the two tones. This is known as *audio FSK* (AFSK).

Two other modes you will run across on shortwave involve the conversion of visual images to a form which can be transmitted using audio tones. One is *facsimile* (FAX), which is a radio version of the system used over telephone lines. "Fax" is mainly used on shortwave to transmit wire service photos or weather charts and maps to ships at sea. However, fax hasn't caught on much among SWLs, primarily due to the cost of receiving units. A new generation of interface devices and software for personal computers to produce fax output on graphics printers promises to change this, however.

The other mode by which visual information is transmitted using audio tones is *slow-scan television*. SSTV converts a still image to a series of audio tones for transmission; at the receiving end, these tones are converted to a form in which they can be displayed on a TV set. Unfortunately, only still pictures can be transmitted, and it takes several seconds to transmit or receive a SSTV image. Moreover, the equipment necessary is often expensive. SSTV is almost exclusively used in the ham radio bands, although a few international broadcasters have experimented with it. You'll be able to recognize SSTV signals as a rapidly changing sequence of audio tones sounding like the sound effects of a video game; the images will usually be prefaced and followed by voice communications.

There are many other types of signals you can hear on the shortwave bands, sounding like "roars," pulses, "beeps," and other esoteric sounds. Unfortunately, the equipment needed to decode these signals is either unavailable or too expensive for average listener. Those modes we've mentioned account for just about everything you'll be able to eavesdrop upon using commonly available equipment.

A Guided Tour of the Shortwave Bands

Strictly speaking, the term *shortwave* refers only to the 1600 to 30,000 kHz (1.6 to 30 MHz) frequency range. However, many so-called "shortwave" receivers actually have broader frequency coverage; a better term for them is *communications* receivers. Frequencies below 540 kHz are known as *longwave*, while the standard AM broadcast band (540 to 1600 kHz) is known as *medium wave*. Frequencies above 30 MHz to approximately 400 MHz are known as *very high frequency* (VHF), while those above 400 MHz are known as *ultra high frequency* (UHF). All of these frequency ranges are divided into bands for specific purposes.

You'll also soon discover that certain bands are useful for DX reception during the day, and other bands are useful for DX reception at night. A good general rule, however, is that frequencies below 10 MHz are best suited for DX reception during evening and night hours, 10 to 15 MHz is useful during the day and night, and 15 to 30 MHz is best for DX during daytime and early evening. Frequencies above 30 MHz produce DX reception only during abnormal, exceptional circumstances. In short, don't expect equal results at the same time throughout the shortwave bands! (The process by which radio signals travel from a transmitting site to listening points is called *propagation* and will be fully discussed in chapter five.)

You'll note that some frequency ranges are referred to as the "80-meter band," "31-meter band," or some similar designation involving meter(s). This is a carry-over from the early days of radio and refers to the *wavelength* of the radio waves in that particular frequency range. This is still a convenient shorthand; "80-meters" is easier to say than "3500 to 4000 kHz," but both mean the same.

The following is a summary of the major frequency bands and what you can hear on them. Keep in mind, however, that while these ranges are generally followed there are some variations in different parts of the world. Moreover, there are some "secondary" uses of most bands in addition to the prime use described. Thus, don't be too surprised to find some other types of signals in each band. In particular, frequencies adjacent to the major international broadcasting bands are often "invaded" by broadcasters.

Below 150 kHz. Most communications receivers don't tune below 150 kHz, although some military surplus receivers do and some receiving *converters* are available to extend the coverage of receivers into this range. The signals in this range are CW and RTTY, and are from military or government stations. One common use for the lower end of this range is for communications with submarines, since those frequencies can penetrate oceans better than any other range. The U.S. Navy's "Omega" submarine navigation system operates on such frequencies as 10.2, 12, and 13.6 kHz. The Soviet Union operates a similar navigation system for its submarine fleet on 15.625 kHz. The U.S. Air Force's Strategic Air Command (SAC) also operates stations in this range, since these frequencies are not affected to a significant degree by atmospheric conditions (and therefore would not be rendered useless by the effects of nuclear explosions in the atmosphere). SAC can be found on 29.5 and 37.2 kHz with coded RTTY transmissions. Other stations you'll find in this range include radio navigation beacons and fixed stations.

150–540 kHz. This is the range most SWLs mean by "longwave." Most activity is CW, although you'll also hear some AM. Many of the stations you'll hear are navigation beacons. Beacons usually do nothing more than continuously repeat their call letters in slow Morse code, and can be heard from 200 to 415 kHz. Unlike other radio stations, longwave beacons do not follow the international allocations for call letters. Instead, their call letters usually suggest their location. For example, beacon CLB on 216 kHz operates from Carolina Beach, North Carolina, and MP transmits from Montauk Point, NY, on 286 kHz. Some of these beacons include aeronautical weather broadcasts in AM; you can hear the call letters of the beacons being repeated in the background using Morse code during the weather broadcasts.

Listeners in eastern North America may be able to hear a few broadcasting stations in Europe in the 155 to 281 kHz range.

The range from 415 to 515 kHz is used for maritime communications in CW. Maritime stations include shipboard stations and shore stations in contact with ships. These maritime stations do use call signs that follow international allocations. Two important frequencies here are 500 and 512 kHz. 500 kHz is an international distress and emergency frequency, and maritime stations worldwide monitor it for calls from ships in trouble. 512 kHz is a *calling* frequency; stations wishing to contact another specific maritime station call it on this frequency and then move to another frequency once contact has been established.

The 515 to 540 kHz range is populated by a handful of miscellaneous beacons, some of which can be heard at considerable distances. These are being phased out. Foreign broadcasting stations can also be heard on frequencies such as 525 and 530 kHz. 530 kHz is also used by low-powered stations near airports and tourist sites to broadcast information to travelers.

Maritime use of longwave has been gradually diminishing, with much traffic now going via satellite and RTTY. Within a few years, maritime traffic may be virtually gone from longwave.

All frequencies in the 150 to 540 kHz range are best received during the night, with best DX receiving conditions found during the autumn and winter months.

540–1600 kHz. This is the standard AM broadcast band familiar to everyone. In North American and most of South America, AM stations operate on 10 kHz channels beginning at 540 kHz and continuing through to 1600 kHz. In the rest of the world, AM stations operate on channels separated by 9 kHz. A few stations in Central and South America operate on frequencies located between the normal 10 kHz channel spacing. Sometimes this is by design, such as YSS in El Salvador on 655 kHz; in other cases, this is the result of faulty transmitter operation. Like longwave, these "medium wave" frequencies are best at night during the fall and winter.

1600–1800 kHz. This is an interesting range, formally allocated to radio navigation but actually a "grab bag" of various radio services. For example, 1610 kHz is, like 530 kHz, used for traveler information. A handful of Morse code beacons similar to those on longwave operate in this range, although these are rapidly disappearing. You can also hear several seismic beacons, usually for oil and gas exploration, producing chirping sounds which some listeners describe as "crickets." Older model cordless telephones are found from 1700 to 1800 kHz, and these can provide hours of *fascinating* listening. A major change to this frequency range will come in the 1990s, as the 1600 to 1700 kHz range will be opened to AM broadcasting in the United States.

1800–2000 kHz. This is known as the 160-meter amateur radio band, and is the lowest frequency range allocated for use by ham radio operators. You'll hear both CW and SSB used here. Most communications you'll hear will involve stations located within a few thousands of miles or less of each other, although some hams have managed to contact all continents and 100 different countries on 160-meters. You'll find reception best on this range during the fall and winter from evening to your local sunrise.

2000–2850 kHz. The major users of this frequency range are maritime stations, along with some land mobile and fixed stations. Most transmissions are in SSB along with a scattering of CW and RTTY. An important frequency is 2182 kHz, which is a voice channel for emergency and distress maritime communications. Ship-to-ship communications can be found on such frequencies as 2082.5, 2638, and 2782 kHz. You'll also find many maritime weather broadcasts in USB from U.S. Coast Guard stations on 2670 kHz. Standard time and frequency stations such as WWV and WWVH can be found on 2500 kHz.

In some tropical areas of the world, the 2300 to 2498 kHz range is allocated to domestic broadcasting and is known as the 120-meter broadcasting band. Many of these stations are low-powered and located in such exotic locations as New Guinea, Indonesia, and Brazil. This range will be humming with activity from around your local sunset to about your local sunrise.

2850–3150 kHz. This band is allocated for aeronautical mobile stations. One common type is the so-called "VOLMET" stations, which broadcast weather conditions for various aeronautical routes. Airlines also use these bands for communications with their aircraft flying international routes. Almost all aeronautical mobile stations use SSB, primarily USB, with some CW used by a few airlines (primarily Aeroflot, the civilian airline of the Soviet Union). You'll find the best reception here from around your local sunset to local sunrise, with fall and winter best.

3150–3400 kHz. The prime allocation here is for fixed stations and some mobile stations. For example, the U.S. Department of the Interior operates a network of stations in the U.S.-administered territories in the Pacific on 3385 kHz. The Federal Emergency Management Agency (FEMA) operates over 60 stations scattered throughout the United States on 3341 kHz (as well as other frequencies). You can also find standard time and frequency station CHU in Ottawa, Canada, on 3330 kHz during the night hours. There is another tropical broadcasting allocation here at 3200 to 3400 kHz known as the 90-meter broadcasting band. Many countries and much rare DX can be found on 90-meters.

3400–3500 kHz. This is another aeronautical mobile band. VOLMET stations at Gander, Newfoundland, Canada, and at MacArthur Airport on New York's Long Island can be heard alternating on 3485 kHz during the evening.

3500–4000 kHz. This is known as the 80-meter amateur radio band, with the 3750 to 4000 kHz range sometimes referred to as the 75-meter amateur radio band. You'll find CW and RTTY from 3500 to 3750 kHz and voice (usually LSB) from 3750 to 4000 kHz. This is a very popular ham band and is crowded during the evening hours. There is also a standard time and frequency station on 3810 kHz; it is HS210A in Guayaquil, Ecuador. Listen for time pulses and Spanish announcements each minute. In Europe and Africa, the 3900 to 4000 kHz is assigned to broadcasting. You can sometimes hear these stations after 0500 UTC if the interference from hams isn't too bad or if a mild ionospheric storm is in progress.

4000–4063 kHz. This is a fixed-station band. In the United States, you're likely to hear ham radio operators who are members of the military affiliate radio system (MARS) operating here.

4063–4438 kHz. This is a very active maritime band. Voice communications are in USB, and much CW and RTTY is also used. 4125 kHz is the international SSB ship calling frequency and is busy throughout the hours of darkness.

4438–4650 kHz. This band is allocated for the fixed and mobile service. One interesting frequency is 4449 kHz, where several U.S. Air Force stations operate in USB.

4650–4750 kHz. This is another aeronautical band. Of particular interest here are several VOLMET stations located in the Soviet Union.

4750–4995 kHz. This is the 60-meter tropical broadcasting band, and you'll find it the best band to hear domestic shortwave broadcasters. Stations in Africa begin to fade in around your local sunset until their sign off at 2300 or 0000 UTC. Among those you can hear are the Ghana Broadcasting Corporation on 4915 kHz, Radio Nigeria on 4990 kHz, and Bamako, Mali on 4835. Most of these same stations can be heard again around 0600 UTC when they sign on (listeners in western North America often find this a better reception time). The evening and night hours are dominated by signals from Latin America, such as Radio Pioneira in Brazil on 5015 kHz, Costa Rica's Radio Reloj on 4832 kHz, La Voz del Cinaruco on

4865 kHz from Colombia, and Radio Quito in Ecuador on 4920 kHz. Many Latin American stations sign off around 0500 or 0600, which is their local midnight. Stations from the Pacific area can be heard from about 0730 UTC to your local sunrise. Among those to try for are Radio Television Malaysia on 4845 around 1200 UTC and the numerous Indonesians, such as the Radio Republik Indonesia station at Ujung Pandang around 1000 UTC on 4753 kHz.

4995–5005 kHz. This range is set aside for standard time and frequency stations worldwide, with most operating on 5000 kHz. During the night, most listeners in North America will hear WWV, Fort Collins, CO, on 5000 kHz. They broadcast an announcement of the time each minute using a man's voice. If you listen carefully after 0500 UTC, you will likely hear another station underneath WWV announcing the time each minute using a woman's voice. That will be WWVH in Kauai, Hawaii. During unusual reception conditions, you can hear other time and frequency stations in China, India, Japan, and the USSR in this range.

5005–5450 kHz. This range is primarily used by fixed and land mobile stations, although some tropical broadcasting stations may be found in the lower part. You'll find SSB, CW, and RTTY used here throughout the evening and night hours.

5450–5730 kHz. The first 30 kHz of this range is shared by fixed stations and aeronautical stations; the remainder is exclusively aeronautical worldwide. Most communications will be in USB and involve VOLMET broadcasts and long-range communications by aircraft flying international routes. Aircraft flying the Caribbean can be found on 5520 and 5550 kHz, while 5598 and 5649 kHz are used by airplanes flying north Atlantic routes between Europe and North America. As with the other bands mentioned so far, you'll find most activity during the evening and night hours.

5730–5950 kHz. This is the range assigned to fixed stations, and numerous stations in SSB, CW, and RTTY can be found here. The U.S. National Weather Service maintains a network of stations using USB on 5923 kHz. The Department of Energy uses 5948 kHz to coordinate shipments of nuclear materials across the country. And the U.S. Air Force and the National Aeronautics and Space Administration (NASA) use 5810 kHz for USB support communications during space shuttle launches. Try tuning this frequency whenever a launch is scheduled, particularly if you're located within 1000 miles of Cape Canaveral.

5950–6200 kHz. This is the lowest frequency range allocated for international broadcasting, and is known as the 49-meter band.You'll find this range packed with AM signals from late afternoon to approximately an hour after your local sunrise. In addition, you may find some signals (although usually weak ones) during the daytime, especially in winter. During the evening hours in North America, this band is full of powerful signals from European international broadcasters. Some interesting signals can be found amid the powerhouses, however, such as Radio Luxembourg on 6090 kHz. Radio Luxembourg is an English-language pop music station patterned after American "hit music" radio. It lets Americans hear some of the newer European artists long before they make it (if ever) to the American music charts. Try for it from your local sunset until a couple of hours later; best reception is often in May or early June. Around 0500 UTC, the European broadcasters start leaving the air as sunrise starts moving across Europe. The reduced interference on 49-meters then lets you hear several stations in South America which operate all night in Spanish. Around 0800 UTC, stations from the Pacific and Asia start to become audible. China, Indonesia, Australia, the USSR, and the Philippines are among the countries you can hear in the hours before and shortly after your local sunrise.

6200–6525 kHz. This range is allocated exclusively throughout the world for maritime communications, and will be busy during the same hours the 49-meter broadcasting band is

FIGURE 2-4

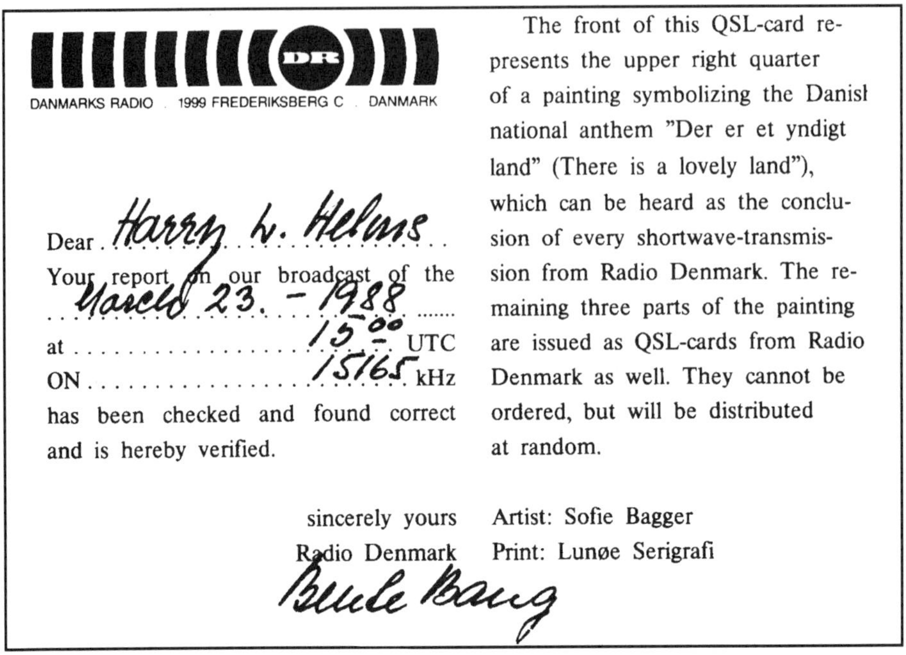

DANMARKS RADIO . 1999 FREDERIKSBERG C . DANMARK

Dear . Harry L. Helms . . . . .
Your report on our broadcast of the
. . March 23. – 1988 . . . .
at . . . . 15^00 UTC
ON . . . . 15165 kHz
has been checked and found correct
and is hereby verified.

sincerely yours
Radio Denmark
Bente Bang

The front of this QSL-card represents the upper right quarter of a painting symbolizing the Danisł national anthem "Der er et yndigt land" (There is a lovely land"), which can be heard as the conclusion of every shortwave-transmission from Radio Denmark. The remaining three parts of the painting are issued as QSL-cards from Radio Denmark as well. They cannot be ordered, but will be distributed at random.

Artist: Sofie Bagger
Print: Lunøe Serigrafi

Radio Denmark can be heard on shortwave, but not from Denmark! It ceased transmissions from Denmark in early 1990 and now is relayed by Radio Norway International. It also stopped sending out QSL cards to listeners, making the one shown here a collector's item.

in use. As with other maritime bands, you'll hear SSB (usually USB) used along with CW and RTTY. 6218.6 kHz will be in use almost constantly for USB communications; this is used for inland (that is, rivers and the like) maritime messages (or "traffic") as well as by the U.S. Coast Guard along the Atlantic and Pacific coasts. Another active USB channel is 6221.6 kHz.

6525–6765 kHz. This is another aeronautical band populated by VOLMET stations, aircraft aloft, airports, and flight operations centers. The U.S. Air Force (USAF) uses such frequencies as 6670, 6683 (often used by Air Force One), 6712, and 6738 kHz for its USB communications. You may hear messages on these frequencies composed of various words and numbers; these are known as "Sky King" or "Foxtrot" broadcasts and are coded instructions to USAF aircraft aloft.

6765–7000 kHz. This is allocated for fixed stations, and is mainly filled with CW, RTTY, and miscellaneous data signals.

7000–7300 kHz. This range is shared by both amateur radio and international broadcasting; it is known as 40-meters. The 7000 to 7100 kHz range is supposedly allocated exclusively to amateur radio worldwide, although a handful of broadcasters use it. 7100 to 7300 kHz is used for international broadcasting in Europe and Asia, while it is assigned to hams in North and South America. The result is that this may be the most crowded band on shortwave, as tuning it during the evening hours will reveal. For international broadcasting, it is very similar to 49-meters in what you can hear and when. Hams use LSB in addition to CW and RTTY. In the United States, 7150 to 7300 kHz is used for voice and SSTV while 7000 to 7150 kHz is used for CW and RTTY. During the day, hams use this band for communications over distances of approximately 1000 miles or less.

7300–8195 kHz. This is assigned worldwide for fixed service, although some international broadcasters can be found at the lower end of the range. You can hear CHU from Canada on 7335 kHz with time announcements in English and French each minute. Interpol stations can be found on 7401 kHz in RTTY. The U.S. Customs Service also uses 7527 kHz in USB to coordinate its activities.

8195–8815 kHz. This is another busy maritime band, with CW used more heavily than USB. 8257 kHz is an international ship-calling frequency for USB; 8291.1 and 8294.2 kHz are other common USB frequencies. 8364 kHz is an international CW frequency for emergency and distress signals.

8815–9040 kHz. This is another aeronautical band. 8825 kHz is used for USB traffic by flights crossing the Atlantic. 8870 kHz is used for VOLMET broadcasts by stations in Gander,

Newfoundland and Long Island, New York. Air Force One has also been heard in USB on 9018 kHz.

9040–9500 kHz. This is allocated to fixed stations, with many using RTTY. You'll also find some international broadcasters scattered through this range.

9500–9900 kHz. This is known as the 31-meter international broadcasting band, and may be the most heavily used band for that purpose. It is also a "transitional" band; best reception is generally during the evening and night, but some stations can usually be heard on this band during the day (especially in winter). During late afternoons, stations from Europe and Africa "fade in" to audibility. During evenings, almost every major European international broadcaster can be found here. These stations leave the air around 0500 to 0600 UTC and are replaced by stations from South America, the Pacific, and Asia. Stations from these areas can often be heard until approximately two to three hours after your local sunrise. Australian stations, such as the Australian Broadcasting Corporation's domestic service outlet near Perth in Western Australia on 9610 kHz, can also be heard well. This station is interesting because it is the most distant station that most listeners in eastern North America will be able to hear.

9900–9995 kHz. This range is allocated worldwide to fixed stations, most of which use RTTY, although broadcasters are intruding here in large numbers.

9995–10005 kHz. This is reserved for standard time and frequency stations. Many of the stations found at 4995 to 5005 kHz can be found here, with WWV heard around the clock in North America.

10005–10100 kHz. This is an aeronautical band largely used by airplanes aloft. Interesting USB frequencies include 10072 and 10075 kHz, which are used by airline companies to communicate with their aircraft on matters unrelated to navigation and safety (communications on these topics are handled by airports).

10100–10150 kHz. This is the 30-meter amateur radio band, which was allocated to amateurs in 1979. Because of the narrowness of this band, operation here is normally restricted to CW and RTTY.

10150–11175 kHz. This allocation is reserved for fixed stations worldwide. In addition to the types of utility communications mentioned previously, this range has many *broadcast feeders*. A broadcast feeder is used by international broadcasters to relay programs from their studios to overseas transmitter sites. Such feeders are gradually being replaced by satellite relays, but a surprising number are still operated by such broadcasters as the Voice of America, Radio Free Europe, and Radio Moscow. You'll recognize feeders since they carry normal broadcast programs, but SSB is used instead of AM. You'll also find numerous Interpol stations on 10295 kHz in RTTY.

FIGURE 2-5

Kol Israel can be easily heard throughout North America.

11175–11400 kHz. This is an aeronautical allocation. 11182 kHz is an active USAF channel, and Russian language VOLMET stations in the Soviet Union can be heard on 11279 kHz. Aeroflot flights enroute between Moscow and Havana can be heard on 11312 kHz—but transmissions are in CW, not SSB.

11400–11650 kHz. This is a fixed station allocation with many RTTY and facsimile stations in this range. International broadcasters are gradually encroaching here as well.

11650–11975 kHz. This is the 25-meter international broadcasting band and often has something interesting around the clock. You'll find most of the major international broadcasters of the world using this band, along with several international broadcasters which do not have programs specifically targeted to North America. For listeners in North America, this is a good band to tune during late mornings and early afternoons for stations in Europe, Africa, and the Middle East. During the evening, this band is often used by major international broadcasters for their North American services. Later in the night, stations from the Pacific and Asia become audible and can be heard until a couple of hours after your local sunrise. One favorite of many SWLs is Radio Television Française d'Outre Mer (RFO) station at Papeete, Tahiti on 11825 kHz. In addition to some authentic "South Seas" music, it may be your best opportunity to hear spoken Tahitian!

11975–12330 kHz. This is assigned to fixed stations worldwide, although international broadcasters are increasingly found in the lower part. Reception characteristics are similar to 25-meters. An interesting frequency is 12216 kHz, where the Federal Emergency Management Agency (FEMA) operates several stations using USB. Several broadcast feeders also operate in this allocation.

12330–13200 kHz. This is a maritime allocation, and is quite busy throughout the day and early evening hours. One very busy USB frequency is 12429.2 kHz; however, most signals on this band are CW or RTTY.

13200–13360 kHz. This is an aeronautical band. Active frequencies used by the USAF for USB communications include 13201, 13204, and 13241 kHz; the latter channel is often used for "Sky King" broadcasts.

13360–13600 kHz. This is allocated to the fixed service. It also lies at the beginning of the portion of the radio spectrum in which best reception takes place during the day and early evening (and, as will be detailed later, during years of high solar activity).

13600–13800 kHz. This is the newest international broadcasting band, known as the 22-meter broadcasting band. Numerous countries have started operations here and the band is often crowded during the afternoon hours.

13800–14000 kHz. This is a fixed station band. Among the stations found here is an emergency network operated by the International Committee of the Red Cross on 13915 and 13397 kHz in USB. The FCC also uses 13990 kHz for RTTY communications between its monitoring stations.

14000–14350 kHz. This is the 20-meter amateur radio band and is usually the best ham band for DX communications. The first 100 kHz is reserved for CW and RTTY, while 14100 to 14350 kHz is used for SSB (usually USB) and SSTV (although U.S. hams can only use 14150 to 14350 kHz for voice communications). 14230 kHz is a very active frequency for SSTV activity. This band is most useful (or "open") for DX communications during the daytime and early evening, although it is often open around the clock during years of high solar activity.

14350–14490 kHz. This is another band for fixed stations, and you'll find CW, RTTY, SSB, and FAX used here. Some stations you can hear include a network of Australian research bases in Antarctica using USB on 14415 kHz and an Interpol network for African nations on 14827 kHz in CW.

14990–15010 kHz. Like 4995 to 5005 and 9995 to 10005 kHz, this is a standard time and frequency allocation; many of the stations you can hear on 10000 kHz can also be heard here.

A good time to listen for a station other than WWV or WWVH is to tune around your local sunrise, when signals from those two stations are most disturbed.

15010–15100 kHz. This is a narrow aeronautical band. A heavily used USAF frequency is 15041 kHz in USB. Listen for the various tactical call signs such as "Morphine," "Tomahawk," and "Overbrook." 15048 kHz is another busy USAF channel, and sometimes Air Force One can be heard there. A few international broadcasters intrude upon this range. The BBC can often be heard during the morning and early afternoon on 15070 kHz, while Iran uses 15084 kHz at the same time for Farsi and Arabic programs.

15100–15600 kHz. This is the 19-meter international

FIGURE 2-6

Radio Nederland has long been a "state of the art" broadcaster. They recently opened new transmitting facilities near Flevo in the Netherlands.

broadcasting band, although some fixed stations can be heard in the upper end of the range. This is a heavily used band during the day and evening hours. It is particularly useful for reception of Asian and Pacific stations during the evening in the summer months; try for RFO in Tahiti on 15170 kHz. Radio Japan, Radio Beijing, and Radio Australia frequently use this band for transmissions during North American evenings. This frequency range, like the 20-meter ham band and all higher frequencies, offers best reception during years of high solar activity (indicated by a high number of sunspots).

15600–16460 kHz. This is a fixed station allocation worldwide. A network of USAF stations in USB can be found on 15632 kHz.

16460–17360 kHz. This broad range is shared by fixed and maritime stations, and is usually humming with activity during daylight. 16523 kHz is an internal USB calling frequency for ships. Other active USB ship channels include 16587, 16590, and 16593 kHz. Station ROT, Moscow, USSR, can be heard in CW on 17130 kHz; this station is operated by the Soviet Navy.

17360–17550 kHz. This is an allocation primarily for fixed stations, although Air Force One has been heard on 17385 kHz in USB.

17550–17900 kHz. This is the 16-meter international broadcasting band and is useful for daytime reception for stations to the east of your location during the morning and early afternoon and for stations to the west of you during late afternoons and early evening. If evening and night reception on 19-meters is good, try for Pacific and Asian stations here as well. Some fixed stations can also be found in the lower end of the band.

17900–18030 kHz. This is allocated to aeronautical use. The USAF uses 17975 kHz for USB communications, and 17925 kHz is reportedly used for emergency USB transmissions during airplane hijackings.

18030–18068 kHz. This narrow range is allocated to the fixed service.

18068–18168 kHz. This is the most recently allocated ham radio band, known as 17-meters. Both CW, RTTY, and USB are used here, with crowding considerably less than on other ham bands.

18168–19990 kHz. This is primarily a fixed service allocation, although the 18780 to 18900 and 19680 to 19800 kHz ranges are shared with the maritime service. A particularly interesting frequency is 19954 kHz, used by the Soviet Salyut and Mir space stations for beacon transmissions. For terrestrial stations, this band is generally useful only during the daytime and may not be able to support DX communications during years of high solar activity.

19990–20010 kHz. This is an allocation for standard time and frequency stations (WWV is found on 20000 kHz) and space satellite transmissions. Transmissions from manned Soyuz vehicles have been reported on 20008 kHz in AM.

20010–21000 kHz. This broad allocation is exclusively for fixed stations worldwide, although some aeronautical stations are found here. 20042 kHz is used for USB communications by the USAF, and 20053 kHz is used by Air Force One. NASA uses 20191 kHz for LSB communications in connection with launch support activities.

21000–21450 kHz. This is the 15-meter amateur radio band, and ranks second only to 20-meters in popularity for DX communications. In fact, during years of high solar activity it can outperform 20-meters for DX. The first 200 kHz are usually occupied by CW and RTTY signals, while the remainder is used for USB and SSTV.

21450–21850 kHz. This is the 13-meter international broadcasting band, and is most useful during the daytime. During years of low solar activity, many broadcasters abandon this range for 16-meters and 19-meters. You'll also find some fixed stations at the upper end of this range.

21850–21870 kHz. This narrow frequency segment is reserved for fixed stations.

21870–22000 kHz. Although assigned for aeronautical stations, several fixed stations are also found in this range.

22000–22855 kHz. This is an active maritime allocation during daytime. 22124 kHz is reserved for USB communications between ships, while 22127 kHz is used by the USCG.

22855–23200 kHz. This band is allocated to fixed stations. During years of low solar activity, you will find very little going on here.

23200–23350 kHz. This is an aeronautical band, with 23337 kHz used for USAF USB traffic.

23350–24890 kHz. This is allocated to fixed stations worldwide. Its usefulness for DX communications depends very heavily on the level of solar activity.

24890–24990 kHz. This is the 12-meter ham radio band, a relatively recent addition authorized in 1979. It is almost exclusively a daytime band, mainly useful in years of high solar activity.

24990–25010 kHz. This range is assigned to standard time and frequency stations, although none operate here at this time.

25010–25550 kHz. This range is assigned to fixed, land mobile, and maritime stations. Many of these stations, particularly in land mobile service, are low-powered units such as those used in taxis, boats, in plants and factories, and the like. However, surprising distances can be covered with low power during years of high solar activity. For example, listeners on the West Coast have reported what seem to be Indonesian and Chinese language taxi dispatching communications here.

25550–25670 kHz. This range is set aside for use by radio astronomy.

25670–26100 kHz. This is the 11-meter broadcasting band. However, it is seldom used even during periods of high solar activity.

26100–28000 kHz. This is set aside for fixed and mobile communications. In the United States and Canada, 26965 to 27405 kHz is used for citizen's band (CB) radio. Illegal two-way communications by "outlaw" CB operators can be found in the 27500—28000 kHz range; many of these will sound similar to ham radio operations, but they are not.

28000–29700 kHz. This is the 10-meter amateur radio band. It is useful for DX communications during years of high sunspot activity and for local communications, with most activity in USB from 28300 to 28500 kHz.

29700–30000 kHz. This range is mainly used for land mobile stations. During periods of good DX reception, don't be surprised to hear dial tones and busy signals here—this range is used for mobile telephone service in Mexico and Central America.

CHAPTER THREE

Selecting a Shortwave Receiver

SELECTING A NEW SHORTWAVE RADIO can be difficult, especially if you've never bought one before. There are a lot of choices, with price tags running from slightly more than $100 to well into the thousands. The specifications are frequently baffling; just what does a phrase like "4 kHz @ -6 dB" or "50 Ω unbalanced input" mean anyway? Like buying a car, it helps to understand a little of what goes on "under the hood"! In this chapter, we'll look at the different types of shortwave radios, how they work, and the meaning and importance of various receiver features and specifications.

Sometimes you'll see shortwave receivers referred to as *communications* receivers. This term usually refers to more advanced receivers capable of receiving various types of signals (AM, CW, SSB, RTTY, and so on) under difficult reception circumstances. Another term you'll often see is *portable* receiver. As the name indicates, this is a radio which can be carried easily and operated from batteries. The distinction between these two types has become blurred in recent years. Many portable receivers now offer performance and features formerly available only in communications models, while many new communications sets are light and compact enough to be easily transported and can operate from large batteries such as those found in automobiles.

Receiver Basics

It really helps to know a little about how a typical shortwave receiver works. This makes it easier to understand the importance of various receiver specifications and to determine which features are important and which are merely the radio equivalent of tail fins on 1950s cars.

All receivers today are *superheterodyne*. This means the received frequency (such as 9500 kHz) is changed to another, fixed frequency (such as 455 kHz or 10.7 MHz) before the radio signal is transformed into an audio signal you can hear. Sometimes a received radio signal is converted to two or even three different fixed frequencies; this is known as *double* or *triple* conversion. If only one fixed frequency is used, it is called *single conversion*. The superheterodyne is used because certain func-

FIGURE 3-1

The Sony ICF-2010 is an example of today's receiver technology. Its features include synchronous detection, memories, and an LCD display.

tions, such as amplifying the received signal, are more easily and efficiently accomplished at a single frequency than over a wide range of frequencies.

Figure 3-2 shows a block diagram of a single-conversion superheterodyne receiver suitable for AM, SSB, or CW reception. Radio signals striking the antenna produce weak electric currents in the antenna. These weak currents are amplified in the *radio frequency (RF) amplifier* section. The amplified signal is then applied to the *mixer* stage. Note that a signal from the *local oscillator* section is also applied to the mixer. As its name implies, the mixer combines the signals from the local oscillator and RF amplifier to produce a new, fixed frequency signal such as 455 kHz. This is known as the *intermediate frequency (IF)*. Regardless of the actual frequency the radio might be tuned to (such as 9500 kHz), the intermediate frequency (such as 455 kHz) remains constant.

FIGURE 3-2

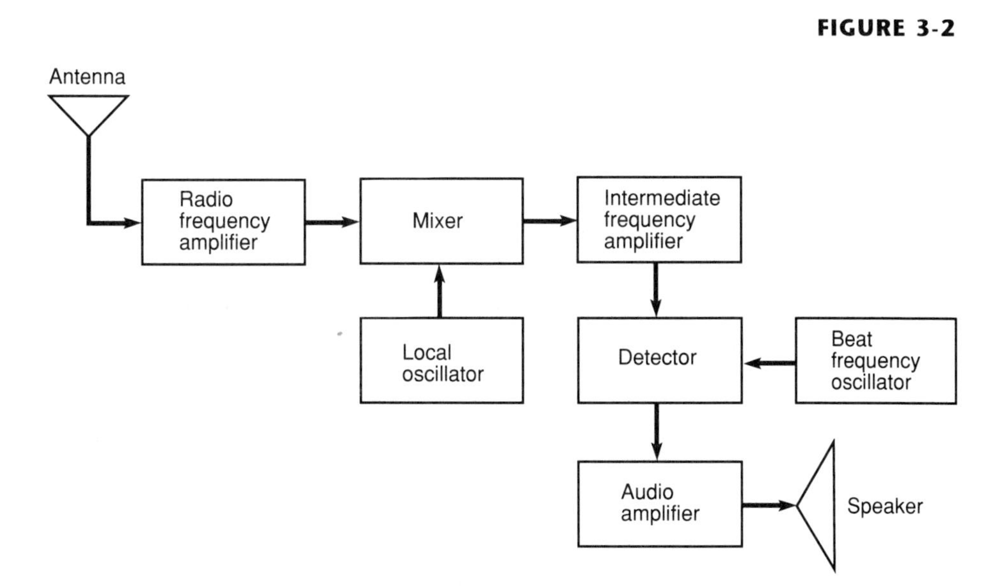

This block diagram shows the basic stages of superheterodyne shortwave radios.

The mixer is normally followed by an *intermediate frequency amplifier*. This stage amplifies the intermediate frequency and feeds it to the *detector*. The detector is the receiver section that converts the intermediate frequency into understandable audio. If the signal is CW or SSB, a *beat frequency oscillator* supplies a replacement carrier to make the signal intelligible. Finally, the *audio amplifier* stage amplifies the audio from the detector so you can hear it through a speaker or headphones.

If double or triple conversion is used, the different mixer and IF amplifier stages are usually "chained" together one after another.

Frequency Coverage

Most shortwave radios are described as *all band*, *all wave*, or *general coverage*, meaning they cover at least from 540 kHz to 30 MHz or so. Many modern SW radios also include coverage down to about 150 kHz, and some also add the FM broadcasting band. An all band or general coverage radio covers the 540 kHz to 30 MHz range without any gaps in its coverage or missing frequencies.

In previous years, many shortwave receivers tuned only certain segments or bands, usually in portions 500 kHz wide. For example, a receiver could tune 40-meters from 7000 to 7500 kHz or 6900 to 7400 kHz, while 31-meters could be tuned from 9500 to 10000 kHz. A receiver could usually tune a dozen or so segments, with users able to determine which segments they wished to tune by inserting the proper crystal for a desired range into the radio's tuning circuitry. Such receivers were very popular among SWLs, and discontinued models such as the Drake SPR-4 and R-4B are still widely used. Similar receivers covering only the amateur radio bands were commonly used by hams.

Almost all shortwave radios made today are general coverage. The only exceptions are a few portables, primarily at the lower end of the price spectrum, which cover only the major international broadcasting bands. Unless price is a very crucial factor in your purchase decision, your best bet is an all band receiver. Such a radio will let you sample and experience the full range of radio listening activity and won't become obsolete if your interests expand or change.

Frequency Readout and Tuning

A few years ago, finding a desired frequency on a shortwave radio was like shooting in the dark—sometimes you hit the target, but more often you missed altogether. On those early sets, moving the tuning dial as little as one inch could cover over 500 kHz in frequency space. Dial markings were often given only every 500 kHz or so, and even then were often inaccurate by over 100 kHz. Some receivers included a fine tuning or "bandspread" dial for the various ham radio bands, but even these were often no more finely calibrated than every 25 kHz or so.

Today, the situation is vastly improved. There's no reason to own a SW radio that doesn't indicate *precisely* which frequency you're tuned to.

Most shortwave radios today have some form of *direct frequency readout*, which displays the frequency the radio is tuned to by using light-emitting diodes (LEDs), fluorescent tubes, or liquid crystal displays (LCDs). Sometimes the desired frequency is reached by turning a dial until the display shows the proper frequency, while in other radios the desired frequency can be entered using a keypad similar to that on a touch dialing telephone; some radios allow both methods to be used. Both can display frequencies with equal accuracy. An LED or fluorescent tube display "glows" when in use and is more readable in darkness or dim light, while an LCD display is visible due to reflected light and can be used only in lit areas or by switching on a light in the receiver. LED and fluorescent tube displays consume more power than LCD displays, making receivers using them best suited for operation from wall outlets rather than batteries.

While LCD, LED, and fluorescent tube displays are a major improvement over old tuning methods, they do have some quirks you should be aware of when comparing receivers. One is the *tuning step* of the receiver. Many contemporary receivers use tuning circuits based upon phase-locked loops (PLLs) and digital electronics, meaning they tune in *steps* or *increments* instead of in a continuous range. A typical radio might have a tuning step of 0.1 kHz (100 Hz). This means the frequency increases or decreases in steps of 0.1 kHz as you turn the tuning dial. Suppose such a radio is tuned to exactly 9500 kHz. As the dial is tuned upward, it tunes 9500.1, 9500.2, 9500.3, 9500.4, 9500.5, and so forth. The radio would be unable to exactly tune such frequencies as 9500.15 or 9500.22 kHz. In practical terms, this is normally no problem. Even "narrow" modes such as CW occupy at least 100 Hz of frequency space, so there is no possibility of missing any signals.

Many communications receivers have multiple tuning rates or *speeds*. This allows you to rapidly tune to a desired frequency using a fast speed, such as a 1 kHz rate, and then fine-tune the signal by switching to a slower rate such as 0.1 or 0.01 kHz.

For most listening, a tuning rate of 0.1 kHz or less will do fine. However, some receivers (mainly less expensive portables) have a tuning increment of 1 kHz or even 5 kHz. Such radios are fine for listening to broadcasters, which mainly operate on frequencies spaced 5 kHz apart from each other, but will be much less useful for listening to hams or utility stations and are almost useless for listening to such modes as CW and RTTY.

The *only* sure way to determine the tuning increments available on a radio is to examine the specifications in the owner's manual or manufacturer's literature and ads. If the frequency display includes a decimal point (for a frequency such as 9500.1 kHz), you can be confident it has a tuning increment of at least

0.1 kHz. If it doesn't, the tuning increment could be 1 or 5 kHz. If you can't be sure, ask the receiver dealer or examine the owner's manual!

Sometimes a tuning increment of 0.1 kHz isn't small enough, such as when tuning RTTY signals or SSB signals for most natural-sounding speech. Some receivers allow tuning in 0.01 kHz (10 Hz) increments, while other receivers include a *receiver incremental tuning (RIT)* control. "RIT" is just a fancy name for a fine tuning control, and lets you tune between increment steps.

One interesting quirk in some receivers is that the *resolution* of the frequency display may be less than the tuning increment. One well-known and popular receiver has a display resolution of 0.1 kHz, but has a slow tuning rate of 0.01 kHz available. However, it can only display frequencies to the nearest 0.1 kHz! This means the receiver can tune such frequencies as 9500.11, 9500.12, and 9500.13 kHz, but the receiver frequency display will still indicate 9500.1 kHz. This isn't a problem in most circumstances, but is something to be aware of.

If there's one feature you absolutely need in a SW radio, it's direct frequency readout. The extra cost is more than offset by the savings in time and frustration. Without direct frequency readout, you'll find it difficult (if not downright impossible) to locate specific frequencies mentioned in this book or other SWLing publications. In the past, many new SWLs abandoned the hobby after a few months because of their inability to find desired stations and frequencies. Direct frequency readout displays make finding a desired frequency as easy as dialing a telephone or selecting a television channel.

If you only remember one thing from this book, remember this: *don't buy a shortwave radio without a direct frequency readout display of some sort*. Life is too short to put up with the problems of trying to listening to shortwave without one.

By the way, the benefits of direct frequency readout are not restricted to new shortwave receivers. External readout devices for popular shortwave radios are available from shortwave equipment suppliers.

Sensitivity

Sensitivity is the term used to describe how well a receiver can respond to faint radio signals and produce audio for you to hear. Sensitivity is provided by the radio frequency amplifier section of a receiver. *Selectivity* is how well the receiver can reject signals on frequencies other than the one you want to listen to. Selectivity is provided by the intermediate frequency amplifier sections of a receiver, and will be discussed in the next section.

Sensitivity is usually defined as the input signal level (that is, the signal delivered from the antenna to the receiver) necessary to give a "signal plus atmospheric noise" output from the receiver at some specified point above the internal noise produced within the receiver itself. The point usually specified is 10 *decibels* (dB). Decibels are based upon the response of the human ear. They are used to express ratios between two power levels and are *logarithmic*. This means a 3 dB increase in power is equal to doubling the power, while a 10 dB increase is equivalent to increasing the power ten times. A single decibel change in a signal is just enough change in the ratio between two signals to be noticeable. (If you turn up the volume of your radio to the point where you can notice that it's louder, you've increased the audio output by approximately one decibel.) Input signals supplied by the antenna are measured in *microvolts* (μV), which are equal to one-millionth of a volt. The smaller the number of microvolts specified, the more sensitive the receiver is.

Now that we know what decibels and microvolts are, what does a receiver sensitivity specification like "0.5 μV for 10 dB S+N/N" mean anyway? A good interpretation might be "a half-microvolt signal fed to the receiver by the antenna will produce an audio output from the receiver in which the radio signal, plus natural atmospheric noise, is ten times stronger than the internal noise produced by the receiver itself." (How strong is the internal noise? To get an idea, disconnect all antennas from the receiver and listen to the noise coming out of the speaker. This internal noise is produced by the random motion of electrons within the radio's components and circuits.)

If you're looking for your first shortwave radio, you usually don't have to worry about sensitivity ratings. This is because some of the biggest advances in receiver technology have involved sensitivity; even simple, inexpensive receivers of today are more sensitive than many professional quality receivers of thirty years ago. Moreover, the trend among international broadcasters has been toward higher transmitter powers and better antennas, which make sensitivity even less important than it was thirty years ago. As a result, almost any currently available shortwave radio will be as sensitive as you need for about 90% of possible listening situations. For special situations, such as DXing, accessory amplifiers to boost the signals from antennas are available (these will be discussed in the next chapter).

There are three other points you should remember about sensitivity ratings and their importance:

- Small differences (0.5 μV or so) in sensitivity have little, if any, effect on what you can or can't hear.
- On lower shortwave frequencies (below 5000 kHz), natural atmospheric noise can be stronger than many stations. Increased sensitivity just means you'll hear the noise better, not that you'll hear more stations!
- A receiver can only process the signal it gets from an antenna. A receiver with average selectivity connected to a good antenna will probably outperform a receiver with good sensitivity connected to an average antenna.

It's not uncommon for many contemporary receivers to be *too* sensitive in some situations. High-powered stations can cause *overloading* in the RF amplifiers of sensitive receivers. This problem often happens in and near international broadcasting bands during hours of peak activity. Symptoms of overloading include "ghost" signals appearing on frequencies where they shouldn't, distorted audio on some stations, and having the audio from a strong station superimposed upon the audio of weaker stations. Overloading happens when the signals from the antenna are too strong for the RF amplifier to handle, much as the sound from speakers gets distorted when you turn the volume of a stereo receiver or amplifier up too high.

An important measure of how well a receiver can handle strong signals is its *dynamic range*. This is the range between a receiver's internal noise level and the signal level at which overloading starts to happen. Dynamic range is measured in decibels, and most communications receivers have a measurement of at least 70 dB. A measurement of over 100 dB is considered excellent.

Sometimes overloading can happen on shortwave due to strong stations on the AM broadcasting band. To prevent this, some manufacturers use *high-pass filters* in their receivers between the antenna and RF amplifier. These filters allow frequencies above 1600 kHz to pass without any effect but greatly reduce the strength of signals below 1600 kHz. Most manufacturers design these filters to be automatically switched in when the receiver is tuned above 1600 kHz so as to not affect reception when listening below 1600 kHz.

A technique used in several earlier shortwave receivers and a few contemporary models is *preselection*. In preselection, the RF amplifier stage has its own tuning control allowing it to be "peaked" for a narrow (usually 500 kHz or less) frequency range. The receiver is far less sensitive to frequencies outside this range. Most receivers using preselection require you to manually tune the RF amplifier for best results; some better-quality SW radios automatically adjust preselector tuning to match the frequency being tuned by the receiver.

The two most common methods of combatting overloading in contemporary receivers is an *RF attenuator* and *RF gain control*. An RF attenuator reduces the sensitivity of a receiver by a fixed amount, such as 10 or 20 dB, or it may allow the sensitivity to be reduced continuously. The reduced sensitivity means that strong signals are less likely to overload the RF amplifier section, but it also means that *all* signal levels—strong as well as weak—are also reduced. The RF gain control allows you to continuously vary the amplification (*gain*) of the RF amplifier stage in much the same way a volume control lets you vary the audio output from a receiver. Both of these controls let you use only the amount of RF amplification necessary to hear a station, reducing the chances of overloading.

Selectivity

You might think sensitivity is the most important specification for a shortwave radio. But it's not—selectivity is much more important with today's crowded shortwave bands. A selective receiver can often produce readable signals in situations where a less selective receiver can't.

In the last chapter, we discussed the concept of signal bandwidth for various modes. Ideally, a receiver's bandwidth should be equal to exactly the bandwidth of the type of signal being received. That's not how things work with real-world receivers, however. Suppose you tune a SW radio to 9500 kHz. It will respond to signals transmitted on 9500 kHz. However, it will also respond to signals on adjacent frequencies such as 9498 and 9502 kHz. It can also respond to signals on 9495 and 9505 kHz, and even to ones on 9490 and 9510 kHz unless selectivity is adequate. An important part of a receiver's selectivity is how well it can reject strong signals on adjacent frequencies. A receiver may be able to reject a weak interfering signal only 2 kHz away from a desired signal but be unable to reject a strong signal 3 kHz away from the desired one.

Selectivity is measured in terms of how well a receiver can reject (or attenuate) an interfering signal located so many Hz or kHz away from a desired frequency range. This desired range is known as a receiver's *bandpass* or *bandwidth*. The degree to which an interfering signal is attenuated is expressed in decibels, and the width of a receiver's bandpass is given as the points at which an interfering signal is reduced by 6 dB (to approximately one-fourth its original strength) and by 60 dB (reduced to approximately 0.0000001% of its original strength).

As an example, let's say you want to receive an AM station on 9500 kHz transmitting a maximum audio frequency of 3 kHz. As we saw in chapter two, this means the AM signal would actually run from 9497 to 9503 kHz, occupying 6 kHz of frequency space. The receiver's bandpass should ideally equal 6 kHz as well. All frequencies below 9497 kHz and above 9503 kHz are unwanted and should be rejected. Suppose the receiver has a rated AM bandpass of "6 kHz at -6 dB down." This means any signal located outside the 9497 to 9503 kHz range will be reduced at least 6 dB. In a similar fashion, typical optimum receiver bandpasses for other modes include 250 to 500 Hz for CW, 1.8 kHz for RTTY, and 2.4 to 2.9 kHz for SSB. Many receivers, particularly communications models, have selectable bandpass settings for different emission modes. Such bandpasses are rated at -6 dB attenuation points in receiver advertising and manufacturer's literature.

What happens if the strength of the desired and interfering signals vary? Suppose the AM signal on 9500 kHz is very strong. In this case, a bandpass of 8, 10, or even 12 kHz can produce excellent listening. Now suppose there is also a strong signal on 9505 kHz. A narrower bandpass, such as 6 kHz or less, would be needed to allow clear reception of 9500 kHz. But if the signal on 9505 kHz is significantly stronger than the one on 9500 kHz, even a 3 kHz filter might not be enough to allow good reception.

The best measure of a receiver's ultimate ability to reject interference is the *shape factor* of its bandpasses. The shape factor is the ratio of a bandpass measured at the -6 dB and -60 dB attenuation points. For example, suppose a bandpass is rated as 3 kHz at 6 dB down ("-6 dB") and 6 kHz at 60 dB down. In this case, the shape factor is 2:1.

An ideal situation would be for a bandpass to have a shape factor of 1:1, but this isn't possible in practice. A shape factor of 2:1 or less is possible and indicates excellent ultimate selectivity. In fact, some professional-quality receivers feature bandpasses having shape factors of 1.5:1 or so. Unfortunately, the shape factor or -60 dB selectivity is often not mentioned in receiver advertising, manufacturer's literature, or receiver reviews appearing in various publications.

Figure 3-3 shows a receiver's bandpass plotted graphically for a selectivity of 3 kHz at 6 dB down and 6 kHz at 60 dB down, giving a shape factor of 2:1. The horizontal axis plots the frequency, with 0 representing the carrier (or *center*) frequency of the desired signal, and the vertical axis represents the amount of attenuation in decibels. You'll note how attenuation is close to zero at the center frequency and increases as you move away from it. The shape of a selectivity curve resembles a skirt, and thus a receiver's ability to reject adjacent QRM is known as *skirt selectivity*. A receiver with a good shape factor is often said to have "tight skirts," and a signal on an adjacent frequency that interferes with a desired one is often said to be "coming in under the skirts."

Selectivity is normally achieved by tuned circuits composed of inductors and capacitors. However, such circuits have difficulty in achieving "tight skirt" selectivity needed in many situations. A good way to greatly improve the shape factor of a receiver's bandpass is to use a *mechanical filter* or *crystal filter* in the receiver's IF amplifier section. Both types of filters are based on the piezoelectric effect—the ability of certain materials to transform electrical energy to mechanical energy and vice-versa. Both types of filters can be designed to pass a certain range of frequencies centered around a receiver's intermediate

FIGURE 3-3

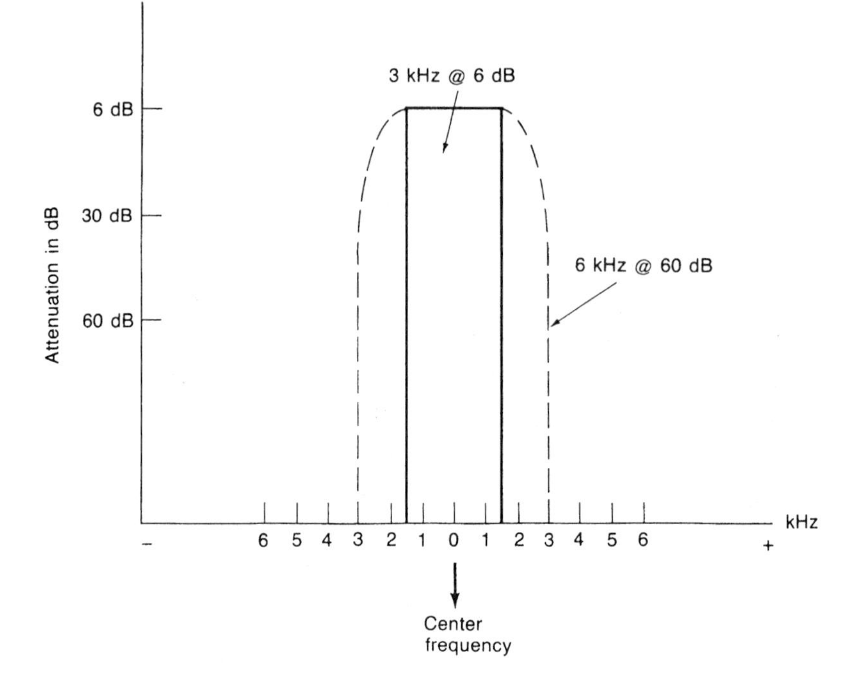

A filter bandpass with a 2:1 shape factor will have selectivity curves like this.

frequency (usually 455 kHz) while rejecting others. Some manufacturers offer crystal filters as optional accessories, usually in bandwidths suitable for SSB or CW reception. Several shortwave equipment dealers will install custom mechanical or crystal filters in the receivers they sell.

Receivers having different bandpasses available will have such controls as a "wide/narrow" switch or a separate bandpass selector control. Other receivers may have different bandpasses that are "tied" into the mode selector switch. That is, setting the mode switch to USB or LSB could select a 2.7 kHz bandpass, while setting it to AM might select a 6 kHz bandpass. Having bandpass selection independent of mode is best, since that allows you to select the most appropriate bandpass for a given listening situation.

A popular innovation in receiver technology is *passband*

FIGURE 3-4

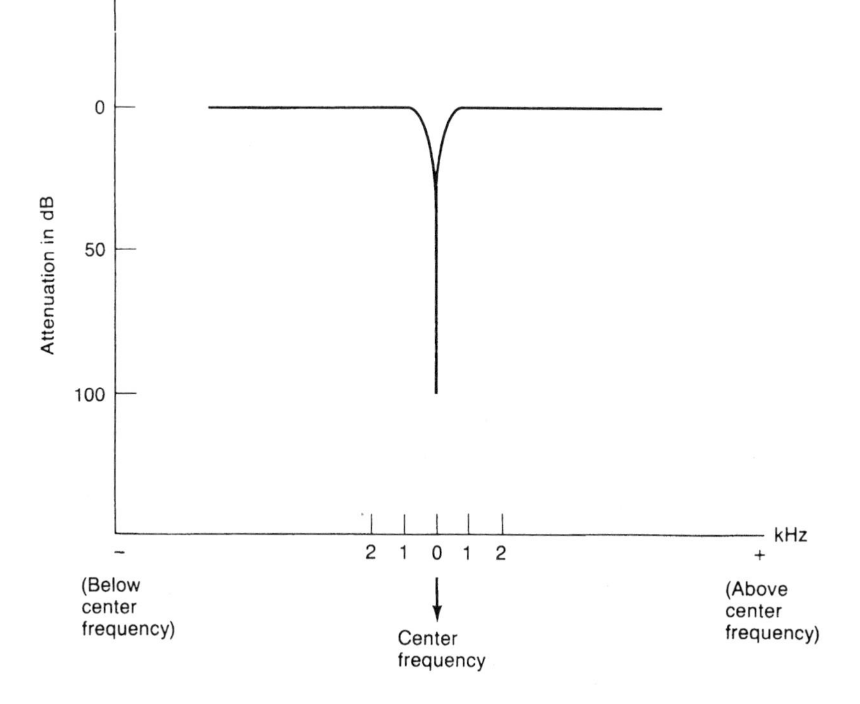

A notch filter has a very sharp rejection at its center frequency.

tuning (PBT), also known as *variable bandwidth tuning* (VBT). PBT/VBT allows the bandpass of a receiver to be continuously varied in much the same way the volume control adjusts the sound coming from a receiver. For example, on a typical receiver with PBT/VBT the selectivity at 6 dB down can be adjusted to any value from 2.7 kHz to only a few hundred Hz. This can be a valuable feature in the hands of an experienced SWL or DXer.

Another valuable selectivity tool is the *notch filter*. Figure 3-4 shows the selectivity curve of a notch filter. Notice how the curve is almost the opposite of the one in figure 3-3? This is because a notch filter works in almost the opposite way a bandpass filter does—the notch filter greatly attenuates any signals on its center frequency. The bandwidth of a notch filter is very narrow, usually a few hundred Hz or less. A notch filter is used to remove a "slice" of the receiver's bandpass where an interfering signal lies. For example, a 2.7 kHz bandpass might be used to receive a SSB signal. Now suppose there is a CW signal, occupying about 200 Hz of space, within the bandpass. Careful adjustment of the notch filter will remove the CW signal and any interference it causes. Using a notch filter inevitably degrades the audio of the signal you want to receive, but in cases of heavy interference a notch filter can mean the difference between having understandable reception and the signal being lost in QRM.

Most notch filters on newer receivers operate only when the receiver is in the CW or SSB mode. Only older tube-type receivers and a handful of modern radios let you use a notch filter in the AM mode. This is unfortunate, since a notch filter can be exceptionally valuable during AM reception. When two AM signals are located close in frequency, the two signals can "beat" against each other and produce a *heterodyne*. A "het" is a piercing, high-pitched whistle that can render either or both signals unlistenable. A notch filter in such cases can eliminate the heterodyne and allow reception of the desired signal.

A "true" notch filter operates in the receiver's IF section, and is sometimes described as an "IF notch filter." A few receivers have what is known as an *audio notch filter*. This is really an advanced tone control, and blocks the audio frequency of the heterodyne whistle. An audio notch filter is not as effective as an IF notch filter.

Some receivers incorporate *audio filters* to aid in selectivity, and numerous models of audio filters are available as accessories. Like audio notch filters, audio filters are simply elaborate tone controls capable of passing a desired range of audio frequencies while reducing or blocking unwanted ones. Audio filters are most effective for CW reception, as the Morse code characters are single-tone and an audio frequency bandwidth of less than 100 Hz is all that is needed. Audio filters are less effective for voice or music reception, and are no substitute for effective selectivity in the receiver's IF circuitry. However, they can be effective in some situations and allow you to tailor the audio from your receiver to your own preferences.

With the shortwave bands becoming increasingly crowded, the selectivity capabilities and options of a receiver are important regardless of whether your listening interests are more toward SWLing or DXing. Selectivity is one area where you do get what you pay for; more expensive receivers have superior selectivity and more options. The ability to choose different bandpasses independent of mode is a valuable feature worth looking for in a receiver.

Noise Limiters and Blankers

One frequently misunderstood receiver control is the *noise limiter* or *noise blanker*. It's also sometimes called the *automatic noise limiter*. Many SWLs are disappointed in the performance of such controls; they are effective against certain types of noise but are ineffective against others. Despite this, a noise blanker or limiter can be valuable in many situations.

Noise limiters and blankers aren't the same thing. A limiter is a simple circuit that "shaves off" the peaks of noise pulses and reduces them to more tolerable levels. A blanker is a more complex circuit that actually silences the receiver during the period of a noise pulse. Both noise limiters and blankers adversely affect the audio quality of a signal, with noise limiters generally the worst offenders.

Noise blankers and limiters are most effective against "pulse" noise such as automobile ignitions, switches, lightning bursts, and similar noises of short duration. They are much less effective against continuous noise sources such as atmospheric static.

A noise limiter often can only be switched on or off. The action of a noise blanker can usually be adjusted. Common adjustments include the degree to which the noise blanker operates, how fast it reacts to noise pulses, and whether it operates on wide bandwidth or narrow bandwidth noise pulses. Another use for noise blankers is in reducing interference from over-the-horizon radar systems operated by the Soviet Union; these systems produce a pattern of signal "spikes" similar to ignition noise.

Automatic Volume and Gain Controls

Many receivers today have some form of *automatic volume control* (AVC) or *automatic gain control* (AGC). These circuits try to maintain the audio output from the radio constant regardless of changes in the strength of the received signal. They sample the level of the received signal and adjust the gain of the RF and IF amplifier sections accordingly. If the received signal weakens, the RF and IF gain is increased; if the received signal increases in strength, the RF and IF gain is reduced.

Many AVC/AGC circuits have *selectable speeds*. The "speed" refers to how quickly the AVC/AGC circuits can respond to changes in the received signal. A fast AVC/AGC is best for SSB and CW because these modes have no carrier and signal strength drops to zero between Morse codes characters or words in SSB. Since signal levels change so quickly, rapid AVC/AGC action is necessary. A slower speed works best for AM, since a carrier is always present and changes in signal level (due to fading, etc.) are more gradual. Using a fast AVC/AGC speed on some AM signals can result in distorted audio.

Almost all AVC/AGC circuits can be switched off. This is because a rapidly fading signal can be distorted by AVC/AGC action regardless of the speed used. In addition, you get maximum receiver sensitivity with the AVC/AGC off.

S-Meters and Signal Indicators

Almost every shortwave radio comes with an *S-meter* or other signal strength indicator prominently displayed on its front panel. Usually such S-meters are calibrated in "S-units" from 1 to 9 with decibels indicated above "S-9" in increments of 20, 40, or 60 dB. Some receivers have LED signal meters, with a stronger signal level lighting more LEDs than weaker signals.

Almost no two different shortwave radios—even two examples of the same model—will give the same S-meter reading for an identical signal. The S-meter or other signal strength indicator is a *relative* indicator of received signal strength, not an absolute measurement such as you get from a thermometer. You'll sometimes see SWLs report in various club bulletins that a signal was "20 dB over S-9" or that it "pinned the meter." While these are colorful descriptions, they don't mean that a different receiver used by the same listeners would indicate the same signal strength. If you have two different radios, you can verify this for yourself by tuning the same station, using the same antenna, and comparing the various S-meter readings.

This doesn't mean an S-meter is worthless. They are useful when using antenna tuners, preamplifiers, or other devices, letting you see when you have everything "peaked" for maximum performance. A visual indication of how rapidly a signal is fading can suggest the proper AVC/AGC speed or even give clues to where the station is located. (Signals traveling over the North Pole and its auroral region often have a rapid, rhythmic fading known as *flutter*.) Just don't take S-meter readings too seriously!

Memories and Multiple Tuning Circuits

One outgrowth of digital electronics has been many receivers incorporating *memories* for storing frequencies. Many of these receivers also have several different ways of tuning frequencies stored in memories or of tuning frequencies outside of those stored in various memories.

You're probably familiar with the "memories" found in telephones, timers, video recorders, and even microwave ovens. In

SW radios, a desired frequency is tuned and then entered into memory (usually by pressing a few buttons in sequence). Often the mode of reception can be entered along with the frequency. Memories are generally numbered, and a desired frequency in a memory can be tuned simply by turning the memory selector to the number representing the desired frequency.

Some receivers come with a *scan* function for memory frequencies or for the normal tuning circuitry. This allows the radio to continuously tune through the memories until the scanning is stopped by the operator or a signal of a certain strength. The time spent monitoring a frequency in memory before scanning to the next varies from receiver to receiver, but is usually quite short. Some receivers allow the scanning function to also operate with the main tuning circuit; lower and upper limit frequencies are specified, and the receiver tunes through the range. The scan rate in such cases is usually equal to the receiver's tuning speed (1 kHz, 0.1 kHz, and so on).

Some receivers allow switching back and forth between the main tuning circuitry and frequencies stored in memory. Other receivers have what are, in effect, two main tuning circuits. These are indicated by such phrases as "dual VFOs" or "VFO A/B" in receiver advertising or front panel labeling. These arrangements permit quickly jumping between two frequencies operating in parallel to compare signal strength, to check the time from WWV, or just to keep track of two different stations.

Memories and various other tuning circuits are handy for both SWLs and DXers. You can store the frequencies of your favorite stations or most wanted DX targets and "flip on" them as easily as you change a television channel.

Beat-Frequency Oscillators and SSB Reception

As we saw in chapter 2, reception of CW and SSB signals requires a beat-frequency oscillator (BFO) circuit. The output of the BFO is fed to the receiver's detector stage. Many communications receivers employ a circuit known as a *product detector*, a combination of a BFO and a special detector circuit for improved reception of SSB and CW.

Simple shortwave radios often have a continuously variable BFO. To receive CW and SSB on such receivers, the BFO is adjusted for the most pleasing CW sound or until SSB is resolved into intelligible speech. More advanced receivers employ fixed-frequency BFOs. Such receivers can be identified by mode selector switches or buttons labeled "CW," "RTTY," "USB," etc. On such receivers, the desired mode is selected and no further BFO tuning is required; however, sometimes the receiver tuning needs to be slightly adjusted for best audio. This is where an RIT control comes in handy.

Even if your prime interest in shortwave is international broadcasting, having fixed selection instead of a tunable BFO will be useful. This is because some AM signals that are weak or suffering heavy interference are best received when tuned as if they were SSB signals. This technique is known as *exalted carrier SSB* (ECSSB) reception.

Exalted Carrier SSB Reception

Reception of AM signals on shortwave can be plagued with many problems. One involves the bandwidth required for an AM signal. If both sidebands and the carrier cannot be received, the signal will be distorted or even unintelligible. Moreover, it is possible (due to the way shortwave signals are propagated) for one sideband to be received before the other one. Even though the difference in the time between reception of the two sidebands can be measured only in picoseconds or less, the delay is enough to "confuse" the receiver's detector stage and produce distortion. Also, the carrier of the received signal must be above a certain level relative to the sidebands in order for the detector to operate properly.

However, it's possible on communications receivers to receive AM signals as if they were SSB signals through exalted-carrier SSB reception. ECSSB gets its name from the fact that the signal from the receiver's BFO is "exalted" over the carrier of the AM signal and replaces it for detection of the signal. In ECSSB, an AM signal is tuned in the usual manner. The receiver's BFO is activated (either by switching it on or choosing USB or LSB) and tuned so that its frequency matches (or is *zero beat* with) the frequency of the AM signal's carrier. By switching to a narrow bandpass normally used for SSB, only one sideband of the AM signal is received. Either the upper or lower sideband of the AM signal can be tuned, but usually the sideband with the least interference is selected.

The term "zero beat" means the AM signal's carrier and the receiver's BFO signal do not beat against each other and produce a heterodyne.To tune the BFO signal to the received signal's carrier, select "USB" or "LSB" on the receiver's mode switch or turn on the receiver's BFO. You'll hear the piercing whistle of a heterodyne. Carefully tune and you'll hear the audio frequency of the whistle drop until it disappears. At that point, the BFO and carrier are "zero beat" with each other.

You might find it necessary to readjust the BFO or receiver tuning every few minutes during ECSSB reception. This is because the BFO and carrier frequency must be within a few Hz of each other, but the BFO frequency in most receivers tends to "drift" over time. More expensive communications receivers have BFOs that are more stable.

Exalted-carrier reception can also be used with wider bandpasses and stronger signals. The replacement carrier generated by the BFO is stronger than the one in any received AM signal. This reduces the effects of fading on the signal and helps the receiver's detector stage do its job better. One setting your author finds effective when tuning the international broadcasting bands is to use the normal AM bandpass but to place the mode selector in USB or LSB (usually USB) and tune for zero beat.

One recent advance in receiver technology has been *synchronous detection*, which can be thought of as "automatic" ECSSB reception. This technique involves receiver circuitry which produces an internal carrier for exalting over an AM signal's carrier and then uses either or both sidebands for detection. While the result is the same as the ECSSB technique, the radio is tuned in the same manner as an ordinary AM radio. There's no need to zero beat or select a sideband to tune for. Some sets using synchronous detection have LEDs or other indicators to show which sideband is being used for detection. While relatively few receivers are currently available using this technology, it holds great promise for improving SW reception and seems certain to be included on more radios in the future.

Learning to Use a Shortwave Radio

Some new SWLs buy the most expensive SW radio they can afford but wind up disappointed when they can't hear certain stations and other things they want to hear. Thinking the receiver is at fault, they sell it to another SWL. A few months later, the original owner of the receiver notices in a SWL club bulletin that the listener who bought the receiver is using it to hear all sorts of rare DX. This isn't because SW radios improve with age like fine wine. It's because *skill* is needed to get the most out of a receiver. This skill involves recognizing reception conditions and being able to adjust the radio's controls and features for best reception under those conditions.

Using a shortwave radio, especially a communications model, is like driving a sports car or using a telescope. A new set of skills is necessary to get the most out of it, and those skills are only acquired with time and effort.

A shortwave radio by itself is nothing but a tool for shortwave reception. Like all tools, you can use it well or use it badly. An experienced SWL using an average receiver will consistently get better reception, particularly of DX signals, than an inexperienced SWL will using a top-of-the-line communications radio.

One thing to understand is that a receiver cannot do everything by itself. A good receiver must be connected to a good antenna system, since a receiver can only work the signal supplied by the antenna. And, as we'll see in chapter 5, radio reception conditions play a major part in determining whether or not you can hear a station. These conditions change from season to season (and even day to day) and set real limits on what it's possible to hear. For example, no matter how good a receiver or antenna one might use, it will be impossible to hear a station from Brazil on 90-meters at noon from a location in central North America.

The best way to use the controls of a shortwave radio varies with reception conditions and the signals you're trying to receive. The correct use of a radio's AVC/AGC and noise blanker can determine whether you're able to understand a signal or not. The same applies to the RF gain control. For example, many SWLs find SSB reception best when they turn the receiver's volume all the way up and use the RF gain control as a "volume" control. This technique doesn't work with all receivers or signals, but is the sort of thing you have to learn from experience and practice.

Selecting the proper bandpass is also something you learn with time. It is possible, for example, to tune a 6 kHz wide AM signal using a 2.7 kHz bandpass filter without resorting to ECSSB. This is done by tuning so you receive the original AM carrier and *part* of one of the sidebands. This results in "clipped" audio and a loss of fidelity, but it can mean the difference between understanding a signal and not being able to identify it at all. Using PBT/VBT or a notch filter for best results is also something that can only be learned with time and practice.

There are other little tricks you learn with practice. Sometimes the best reception of an AM signal can be had by tuning a kHz or so *off* the carrier frequency. On the lower frequencies, such as below 8 MHz, you can often improve readability by *reducing* the RF gain of your receiver. Each receiver has its own quirks, and it will take time to learn just what yours is—and isn't—capable of.

How do you learn these tricks? Start off with the louder stations on shortwave or the AM broadcast band. Play with the various controls of your radio and see what effect they have on the signals. Don't be afraid to try different combinations and to set controls to different positions than those the owner's manual suggests—after all, you can't break the radio by doing so and (I hope this isn't too big a secret!) the people who write the instruction manuals that come with shortwave radios often don't really understand how to best use the radios. Experiment, for example, with listening to a strong local AM station using the USB and LSB modes of your radio. Trying adjusting the RF gain controls and using different selectivity bandwidths. It's a lot easier to learn on such loud signals than it is with weak signals on crowded SW bands!

When you tune shortwave, try tuning much less rapidly than you would an ordinary AM or FM radio. The same turn of the tuning knob that might move you past a half-dozen local AM or FM stations can zip you past several dozen shortwave stations!

You have to develop your SWL "ears." It takes experience to recognize weak signals buried in QRM that might be improved by using a narrower bandpass or ECSSB. DX signals, by definition, are not powerful, easily noticed signals. Less experienced SWLs tune right past these, while more experienced listeners realize that something is buried down in the "mud."

Don't expect to hear everything in this book in one night or in one week. Don't blame your radio if you can't. Practice using your receiver under different receiving conditions with different signals. And, above all, don't get discouraged. We all fell down some when we learned to walk, but there was no other way to learn how to walk. Eventually, we got the hang of it. The same applies to SWLing. The skills will surely come, and soon you'll be able to hear everything mentioned in this book and much more!

Which Receiver to Choose?

There is no single best receiver for everyone. The one that's best for you depends upon your interests and level of experience.

If you're starting out in SWLing, a simple portable SW radio with direct frequency readout and SSB/CW capabilities should do fine. Such a radio will let you sample what shortwave radio has to offer as well as gain experience and define your listening interests more precisely. If your main listening interest turns out to be listening to major international broadcasters, such a receiver will likely be all you ever need. If, on the other hand, SWLing fails to hold your interest, you haven't made a major investment and you still have a radio for travel or emergency use.

Reviews of shortwave radios can be found in magazines such as *Popular Communications* and *Monitoring Times*. SWL club bulletins and their special publications also have receiver reviews, although these are uneven in quality (many reviewers for SWL club bulletins are frankly not technically competent enough or sufficiently experienced to evaluate some receivers). Perhaps the best reviews of all are prepared by Larry Magne and published in each annual edition of *Passport to World Band Radio*. Larry also prepares a series of "white papers" reviewing various popular receivers which are available. Addresses for these and other useful publications are in the appendix.

One thing to beware of are reviews which appear in general interest consumer publications. The few that I've seen have been appalling. The reviewers for such publications inevitably lack the necessary familiarity with shortwave radio to intelligently review a receiver. One memorable review I remember castigated a radio for poor performance on frequencies below 10 MHz; it turned out the reviewer had only used the radio in the early afternoon on a summer day!

If the choices facing you when selecting a receiver seem daunting, you can at least take comfort in the fact that your dollars buy more and better performance than ever before. While making the *best* selection can be a problem, it's difficult to make a *bad* choice today!

CHAPTER FOUR

Antennas and Accessories

ANTENNAS may be the most misunderstood topic among SWLs. A quick glance through various SW club bulletins and commercial publications will show there's no shortage of information (and misinformation!) about antennas. Yet somehow many people wind up with either too much antenna or one that's inadequate for their needs.

A similar situation exists with accessories. Properly selected (and used) accessories can greatly enhance the "DX-catching" ability of a receiver. But poorly chosen or incorrectly used accessories can easily *decrease* the performance of even the best SW receiver.

With both antennas and accessories, it's often true that the most expensive or complex solutions aren't the best. A little knowledge about the two lets you decide on cost-effective enhancements for your shortwave radio. Let's take a look at them in more detail.

Some Antenna Theory

The subject of antenna theory can be complex—just look at any book on antennas published for ham radio operators or communications engineers! That's because those users of antennas are usually concerned with *transmitting* antennas. The wrong antenna when transmitting can damage station equipment. Fortunately for us SWLs, the requirements for receiving antennas are less demanding. The only penalty for using a "wrong" antenna is poor reception. (Of course, any receiving antenna must be installed and used with safety considerations in mind. We'll talk about these later.) You can install and use a shortwave receiving antenna in places and situations where you couldn't do the same with a transmitting antenna. As long as you prevent stray, unwanted electrical charges (lightning, contact with power lines, and the like) from entering your receiver via the antenna, you can connect whatever type of antenna you want to your receiver without worrying about damaging your radio.

The basic principles behind antennas are simple. Radio waves traveling through space strike an antenna and cause very feeble electric currents to be produced (or *induced*) in the antenna. These faint currents then flow from the antenna to the receiver. This simple description of how an antenna works intuitively suggests that a larger or longer antenna works better than a smaller or shorter antenna, since the bigger antenna would logically seem to be able to "capture" more of the energy present in radio waves. It also suggests that an antenna should be clear of metallic obstructions which could absorb radio waves before they reach the antenna. The second observation is always correct; every antenna should be "in the clear" as much as possible. The first observation is sometimes correct, but not always. Depending on the frequency being tuned, a shorter, smaller antenna may outperform a longer and larger one.

The reason why bigger is not always better with shortwave antennas is that an antenna gives best performance when it is *resonant* at the frequency being tuned by the receiver. Whether or not an antenna is resonant depends on the *wavelength* of the radio signal being tuned. All radio signals travel as series of waves having positive and negative peaks, as shown in figure 4-1. The waveform of a radio wave is known as a *sine* wave. The wavelength of a radio wave is defined as the distance between two positive peaks of the wave. As we noted back in chapter 2, wavelength is measured in meters. Thus, the wavelength of radio waves in the 3500 to 4000 kHz range is between 80 to 75 meters long. As signals increase in frequency, their wavelength decreases.

An antenna is resonant at a given frequency when its length is such that it takes a complete radio wave cycle for an electric charge to travel from one end of the antenna to the other and back again. This is defined as the *electrical* length of the antenna, and can be much different from the actual physical length of the antenna. In fact, the physical length of the antenna is only one factor in determining the electrical length of the antenna. It's possible to use coils of wire (called *inductors*) and tuning capacitors to alter the length of time it takes an electrical charge to travel across the antenna and back. This means a resonant antenna can be longer or shorter than a physical half-wavelength, and that a single antenna length (such as 100 feet) can be "tuned" to resonance on a number of frequencies by using a tuning device constructed from variable inductors and capacitors.

FIGURE 4-1

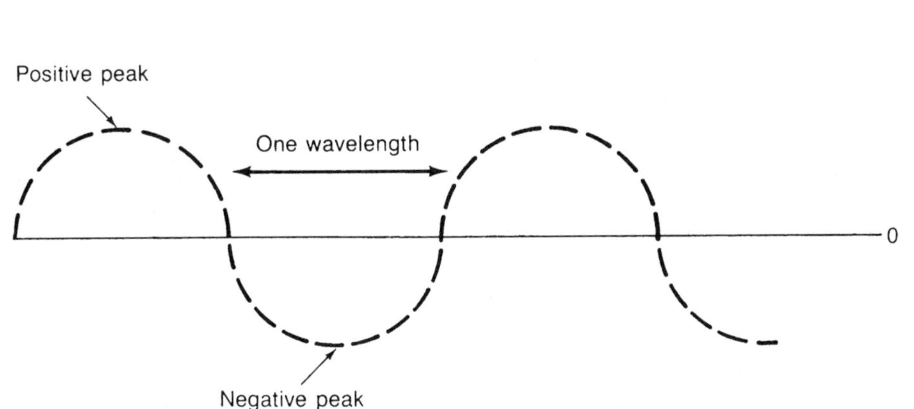

Wavelength is defined as the distance between the positive peaks of a radio wave.

Why is a resonant antenna important? Because maximum energy is transferred from the antenna to your receiver at resonance. This involves *impedance*. Impedance can be thought of as a circuit or device's opposition to the flow of a current. When two electrical circuits or devices are connected, the maximum power transfer takes place when the impedances of the connection points are equal. In our case, this means the impedance of an antenna should equal the impedance found at the receiver's antenna input connections. Impedance is measured in *ohms*, represented by the symbol Ω.

A resonant antenna for a given frequency that is one-half wavelength long has an impedance in the range of approximately 35 to 70 Ω. For this reason, almost all SW receivers have an antenna input impedance of 50 Ω. This is an impedance value near the middle of typical resonant antenna impedances, and any differences in values between the receiver and antenna aren't significant in the real world. Maximum power is transferred between the receiver and antenna, and the receiver has the most signal available to work with.

Untuned, nonresonant antennas have impedance values which can vary widely but are usually in the neighborhood of several hundred ohms or even higher. When such an antenna is connected to a receiver's 50-Ω antenna input, the mismatch is severe, and much of the signal developed in the antenna cannot be transferred to the receiver and is "lost." To partially compensate for this, some receivers have a *high-impedance* antenna input (sometimes labeled *high-Ω* or *high-Z*) in addition to the 50 Ω antenna input. These high-impedance inputs have impedances of several hundred ohms and are better able to match untuned antennas. However, there's usually a remaining impedance mismatch and some loss of signal. This can be verified by using the same random-length antenna and connecting it without a tuner to a receiver's high-impedance input and then switching it through a tuner to the 50-Ω input. A significant improvement in signal strength can be obtained using the tuner through the 50-Ω input. While the difference isn't important in most cases, it can mean the difference between hearing a weak DX station or just hearing noise.

The 50-Ω connector on receivers is designed to accept a coaxial cable plug, while high-impedance connectors are usually of the "spring" or screw terminal type designed for direct connection to random wires. There is usually no problem in using antenna tuners, since most have an input connection for random wire (such as screw terminals) and an output for coaxial cable. A few tuners, however, use "phono" plug connectors (often found on stereo equipment) as a cost-saving measure. Adapter plugs for such situations are available from electronics parts and shortwave equipment dealers.

How Much Antenna Do You Need?

Probably not as much as you think. The temptation for many SWLs is to erect as large and elaborate an antenna system as possible, but most of them could get satisfactory results with something much simpler and cheaper.

In the last chapter, we noted how most shortwave receivers have more than enough sensitivity and that major international broadcasters are using high power transmitters and efficient antennas. These two factors mean you can get by using a very simple (i.e., crummy) antenna if you're mainly interested in listening to international broadcasters and other easily heard stations.

Almost all portable receivers can be equipped with a telescoping "whip" antenna. In most cases, this will give perfectly satisfactory reception for everything but rare DX stations. In fact, the addition of an outside antenna can produce overloading in many such receivers. If you're using a portable receiver, adding an outside antenna to it might be more trouble and expense than it's worth!

More expensive communications receivers can also give

satisfactory performance in most situations using simple antennas. However, this means a lot of the performance and potential of the receiver would be wasted because of an inadequate antenna. If you're going to spend hundreds or thousands of dollars on a premium receiver, you might as well spend another $50 or so to install an antenna that gives your receiver as much signal as possible.

Fortunately, a good antenna system for 1600 kHz to 30 MHz is simple to make and install, as we'll see in the next section.

A Simple 1.6 to 30 MHz Outdoor Antenna

Figure 4-2 shows an antenna system capable of giving excellent results from 1600 kHz to 30 MHz. You can find all the parts, except possibly the antenna tuner, at a Radio Shack store. All components, including the tuner, can be found at ham radio and shortwave equipment dealers. You'll need two antenna insulators (either the "egg" or "dog bone" type), nonmetallic support rope, insulated wire to connect the antenna to the receiver (called "lead-in"), and bare copper wire. The best bare copper wire is the stranded variety in #12 or #14 gauge; solid wire tends to "kink" more than stranded. Insulated wire can be substituted for bare copper wire, since radio waves penetrate plastic insulation easily and lose no strength.

Above all, you have to keep safety in mind when installing any outdoor antenna! Carefully survey the area in which you plan to install the antenna, including the path the lead-in wire will take. Make certain there is no possibility of any contact WHATSOEVER between any part of the antenna system and power lines, lighting fixtures, telephone lines, air conditioning units, transformers, or other voltage sources! Each year in the United States, several people are electrocuted while attempting to install various types of antennas. Almost all such deaths could have been prevented by careful advance planning. Fortunately, almost no deaths result from attempts to install SWL antennas. Let's keep it that way!

FIGURE 4-2

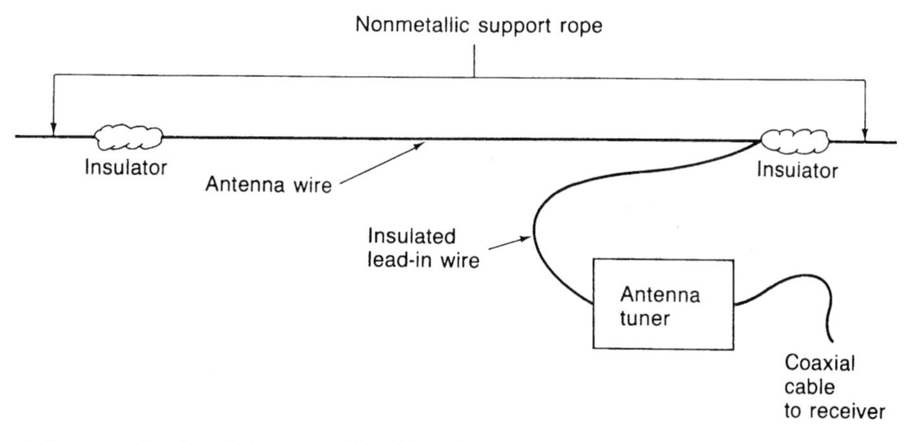

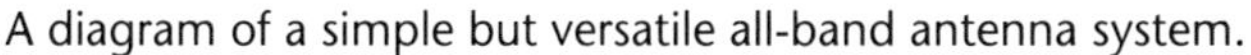
A diagram of a simple but versatile all-band antenna system.

Antenna insulators come with two holes in them. Pass the nonmetallic support rope through one hole of each insulator and secure the support rope tightly to each insulator. The support ropes can be attached to various supports such as trees, poles, sides of buildings, and the like. (DON'T attach to an electric power or telephone pole, however.) The other hole on each insulator is where the wire making up the antenna itself is connected. Approximately two feet of the antenna wire should be passed through this hole and looped through it a few times. The remaining wire is then twisted around the antenna wire or insulator. The length of the antenna wire itself is not critical, but a length of 50 to 100 feet is a good choice. You won't notice a major difference in reception using shorter or longer wire lengths unless the length is significantly shorter (25 feet or less) or longer (over approximately 250 feet or more).

The one crucial connection involves attaching the insulated lead-in wire to the antenna wire. The attachment can be made wherever it's most convenient for you—at either end of the antenna or even in the middle. Note that plain insulated wire is used instead of shielded cable (such as coaxial cable), as the lead-in wire itself absorbs additional energy from radio signals. The lead-in wire's connection to the antenna wire should be soldered, not merely wrapped or mechanically connected (as with an "alligator" spring clip). This is because a mechanical connection will quickly become loose or deteriorate, causing a high resistance connection (or even a broken connection) between the lead-in and antenna wires and making the antenna much less effective.

You should make any solder connections before installing the antenna. A good solder connection consists of a strong mechanical connection soldered with enough heat and the right kind of solder. To make the mechanical connection, select the area on the bare wire you want to connect the lead-in wire to. Using fine sandpaper, gently rub a two inch long area at that point until the copper is bright and shines like a new copper cent. Remove any stray particles and grit from the area with a damp cloth. Remove about four inches of insulation from the lead-in wire and repeat the process on the exposed wire. Remember to use a fine sandpaper and rub gently since the wire can be damaged if you're too rough.

Wrap the lead-in wire tightly around the antenna wire. A good method is to leave approximately one-half inch of stripped lead-in wire "below" the point where the lead-in first makes contact with the antenna wire and wrap any remaining exposed lead-in wire around this section. Some SWLs prefer to wind by looping three or four turns of exposed lead-in around the half-inch section. They then resume looping the exposed lead-in around the antenna wire and repeat the process as necessary. A little practice will soon show which method works best for you. The only requirement is that the resulting mechanical connection be both tight and secure.

The connection must be soldered using only rosin core solder; acid core solder used for metal work will corrode electrical connections. Use a soldering iron of about 30 to 400 watts, although a higher wattage iron or soldering gun might be necessary if you're making the connection outdoors on a cool day. To make a good solder connection, apply heat to the connection to be soldered rather than to the solder itself. Don't melt solder with the iron and then let it drip or flow over the connection. Instead, apply the iron directly to the lead-in/antenna wires. It's best to heat the "bottom" of the connection for several seconds and then apply solder to the "top" of the connection. Keep applying solder until the entire connection is coated with solder and then remove the iron. Don't touch the connection until it's cooled for a minute or so—this is necessary to let the solder "set" properly. If you've done a good soldering job, the result will be a connection covered by bright, shining solder. Dull, pitted solder indicates a poor connection that won't conduct properly and will probably come loose. If it is dull or pitted, apply more heat to the connection until the solder "reflows." After the connection has cooled, wrap it in weather-resistant tape such as electrical tape.

Unlike some antennas, this system will show few (if any) directional characteristics, so the direction in which you install it is unimportant. The prime consideration in installing this antenna is safety. Once that is satisfied, the only real criterion in deciding the location of the antenna is convenience in installation. Using the nonmetallic support ropes, hang the antenna between any two supports (but NEVER power or utility poles). Typical supports could be your house, garage, utility building, trees, or non-metallic poles such as surplus wood telephone poles.

The lead-in is brought into the house for connection to the antenna tuner. Most lead-in wire can be easily slipped under a closed window without difficulty. Some packaged antenna kits for SWLs include what is known as a "window feedthrough" or similar term. This is a short (usually less than a foot) length of flat wire, similar to the "twinlead" cable often used with TV antennas, that has metal connectors on each end. The idea behind this is to slip the window feed-through under a window sill and close the window. The lead-in is then connected to the metal connector outside the window, and the antenna tuner is attached to the connector inside the window. All connections should be soldered. The value of this item is dubious; it's hard to imagine a situation where the window feed-through could fit under a window but the lead-in wire couldn't. It's a real mystery why some manufacturers continue to include this item in their antenna packages; it's more trouble than it's worth.

Most antenna tuners are designed for amateur radio purposes. However, these all work well for SWLs. There are also antenna tuners designed specifically for SWLs, and these are often more cost-effective than those intended for amateur use. The best type of tuner is what is described as a "random wire" tuner. Happily, this type is inexpensive and easy to use. A random wire tuner usually has only two controls, labeled inductance and capacitance. The "inductance" control moves in steps (like the channel selector of a television) while the "capacitance" control is continuously variable, like a volume control. On the rear of the unit is a pair of coaxial input connectors (described as "SO-239" connectors in manufacturer literature). The antenna lead-in is connected to one SO-239 and the other is used to connect the tuner to the receiver using a short length of coaxial cable with input (PL-259) connectors. Connecting the lead-in to the tuner is simple. Remove about three or four inches of insulation from the end of the lead-in and "fold" the wire together to make a "plug." The plug is then inserted into the SO-239 antenna input.

Using the tuner is simple. The "inductance" control is set until the station being tuned is loudest or the background noise is greatest. The "capacitance" control is then rotated for peak signal strength or maximum background noise. As an analogy, think of the "inductance" control as "coarse tuning" and the "capacitance" control as "fine tuning." For best results, you'll have to re-adjust these control every few hundred kHz as you tune through the shortwave bands.

FIGURE 4-3

The Grove Minituner is an example of a simple but effective antenna tuner suitable for use with the antenna shown in figure 4-2.

This antenna is very easy to construct, but will give you years of satisfactory service if carefully built and installed. If you have the space for it, this antenna is probably the best choice—all things considered—for SWLing and DXing from 1600 kHz to 30 MHz.

Coaxial Cable, Grounds, and Lightning Arrestors

The uses and limitations of coaxial cable and lightning arrestors are often poorly understood, and SWLs often misuse them. In the former case, you lose only money and performance. In the latter case, you could lose your life.

You might wonder why coaxial cable wasn't used as the lead-in for the antenna we just discussed. Coaxial cable was originally developed for transmitting applications. Its construction involves a center conducting wire surrounded by foam or plastic insulation. In turn, this insulation is surrounded by a metal braid known as a *shield*. This metal braid is covered by an outer layer of rubber or plastic insulation. "Coax" comes in different sizes and types, and is rated by its ability to handle transmitter power and how much of that power is lost in the coax itself before reaching the antenna. Common types of coax include RG-58 A/U, RG-8U, and RG-213. The impedance of these types is 50 Ω.

The principal advantage claimed for coax is that it prevents noise and unwanted signal pickup because it is "shielded" by the metal braid, which prevents radio signals from entering or leaving the coax. In addition, it's claimed that coax prevents the loss of precious signal strength between the antenna and receiver. Actually, these claims are true when coax is used with certain antenna types (such as the dipole) and receivers. But when used with other antenna types, such as random wires, coax's advantages are wasted.

Coax must be properly *terminated.* This means its metal shield must either be connected to half of a "balanced" antenna (such as a dipole) or to *ground.* In electronics, "ground" does not necessarily mean the earth itself; it means a point of zero voltage (which the earth is) to which a circuit can be connected without affecting its operation. If coax is not properly terminated, the shield has no effect and can even act as a random wire antenna!

What would happen if you tried to use coax for lead-in with the antenna in the previous section? If the coaxial cable is attached to the tuner through a PL-259 connector, the metal braid of the coax will act as part of the antenna wire instead of a shield—the exact thing that happens if you used insulated wire instead of coax for the lead-in. If the center wire of the coax is attached to the tuner without using a PL-259, the metal braid simply "floats" electrically and does nothing—it delivers no signal to the tuner, and it doesn't prevent unwanted signals from penetrating the coax.

Despite this, you might hear of SWLs who swear the performance of their random wire antenna increased markedly when they switched from ordinary lead-in wire to coax. This is because those SWLs finally made a proper connection to the antenna itself when switching to coax. Many SWL equipment suppliers stock coax and recommend it to their customers. There's a good reason—called profit—but it has nothing to do with performance. If you're using a random wire antenna, the only coax you need is a short length to connect the tuner and receiver. Otherwise, forget about coax unless you're using an antenna whose design requires its use.

Grounds are another source of confusion. For years, SWLs have been urged to "properly ground" their receivers by running an insulated wire from their receiver's ground terminal to a cold water pipe or metal rod driven into the earth. Even contemporary SW receiver manuals recommend this practice. SWL "folk wisdom" holds that grounding a receiver increases its performance and protects the receiver from lightning.

It ain't so, folks.

Grounding was often necessary back in the days of vacuum tube receivers, as stray voltages were often found on the chassises of those sets. It is still true with a few modern receivers, such as the Japan Radio Corporation's NRD-515, designed primarily for commercial applications. But most of today's SW radios are adequately grounded through the AC power line. Adding a ground wire to such radios not only doesn't enhance reception, it can actually bring in more electrical noise and hurt reception more than it helps!

Why? A wire connected to a cold water pipe or ground rod will conduct direct current or very low frequency radio energy (a few hundred Hz or less) well, and does a good job of draining away stray voltages from the receiver chassis. But our old friend impedance comes into play as the frequency of the radio energy increases. At shortwave frequencies, the impedance of the wire will be so high that the cold water pipe or ground rod is effectively "disconnected" from the circuit and does nothing. If anything, the wire acts like another antenna and can "de-tune" the antenna system! (To get a good ground at radio frequencies, large flattened copper braid, known as "grounding strap," is needed.)

As with coax, you'll find SWLs who swear that receiver performance increases when they added a ground. (Even the technical columnist for a monthly publication, who should know better, continues to propagate this myth.) In such cases, what has happened is that the SWLs have changed some other aspect of their antenna system without realizing it—or the ground wire itself acts like another antenna. If you still think a ground connection is valuable, try a little experiment: install a ground like your receiver's manual instructs, and listen to various stations with the ground connected and then disconnected. Notice any change at all? That's why your author hasn't used a ground connection in years!

Far more dangerous is the myth that a ground connection can protect your antenna, receiver, or home against lightning strikes. Let's be blunt: *Nothing can protect a receiver against a lightning strike entering from the antenna. NOTHING. There is only one way to protect a receiver against lightning and that is to disconnect the antenna and unplug the AC line cord during thunderstorms or if it is to be left unattended during thunderstorm.* Anyone who tries to tell you differently doesn't know what they're talking about.

There's a category of devices known as "lightning arrestors," and these are dangerously misnamed. These are designed either to short a voltage surge to ground or "pop" like a fuse when a voltage surge is detected. "Lightning arrestors" can be valuable in protecting the circuitry of a modern receiver from damage due to static electricity which builds up on an antenna. That's the main benefit of them—and it's a significant one. But they're worthless in preventing damage from lightning.

If a thunderstorm threatens, unplug your receiver from the AC line and disconnect it from your antenna. Some listeners go so far as to remove the lead-in from inside their homes and hang it outside from hooks or other supports when they're not listening. A lot of trouble? Sure—but not as much trouble as having to buy a new receiver to replace one destroyed by lightning.

Other Types of Antennas

We've mentioned the *dipole* a couple of times. This is a good choice if you're interested in listening to a specific narrow frequency range, such as 500 kHz.

A dipole is a wire that is approximately one-half wavelength long at the frequencies you want to listen to. It is divided in the middle by a center insulator. Coax is used as lead-in, and the center conductor is connected to one "half" of the dipole while the coax shield is connected to the other. The dipole has an impedance of about 75 Ω and an antenna tuner is normally not needed. Dipoles are not much more difficult to build a[illegible] than the random wire antenna previously discussed.

The big drawback to dipoles is that their performance falls off greatly once you tune away from the frequency range they're "cut" for. Thus, a dipole is best once you have enough SWLing experience to know those bands you're most interested in listening to. Full details on dipole theory and construction can be found in various amateur radio antenna books.

A variation is the *trap dipole*, which covers several different bands with a single antenna. "Traps" are inductors (coils of wire) enclosed in weatherproof plastic. A trap dipole has a center insulator and two sections through which the traps are spaced. These traps make the dipole resonant on several different bands, usually the major international broadcasting bands and 60 meters. These are usually more compact than an equivalent "full size" dipole and are easy to build and install. However, they are often expensive and do not perform significantly better than random wire antennas.

Another antenna type used by some SWLs is the *vertical*,

which can be constructed or purchased commercially. The typical vertical is a metal rod one-quarter wavelength long mounted above a set of *radial wires* extending outward from the point where the vertical rod is mounted. The radial wires may be buried underground if the vertical is ground mounted or extended outward with supports if the antenna is mounted on the roof of a building. Like dipoles, verticals can be electrically shortened by adding traps to the vertical rod. Verticals have been favored by ham radio operators interested in DX, since a vertical responds well to signals arriving at a low angle above the horizon (this happens to be the angle from which many DX signals arrive). Another advantage is that verticals can often fit into less space than dipoles.

Verticals do have shortcomings, however. An adequate radial system is essential for proper operation; too few radials can mean poor performance. Many verticals need guy wires to keep them from being blown over by winds.

Verticals have much potential for serious DXing. Unfortunately, commercially available verticals don't cover such prime DX bands at 90 and 60 meters and are expensive for the performance they deliver.

Antennas for Limited Space and Difficult Situations

If you find yourself living in a condominium, apartment, or other area which prohibits the erection of outside antennas, you've got company. Fortunately, living in such residences is no bar to enjoying SWLing.

If you live in a wood or masonary structure without a metal frame, you can get by fine with an indoor antenna. Radio waves penetrate wood easily, and a single brick or stucco wall doesn't significantly weaken signals. The situation changes drastically if you live in a building constructed with steel and reinforced concrete. Such buildings absorb a lot of radio energy, and antennas inside them are virtually useless. The same thing applies to antennas inside mobile homes. Homes and other structures with aluminum siding may or may not present similar problems; it all depends on whether there is another path for radio energy to enter the structure (such as through a nonmetallic roof).

The traditional installation for an indoor antenna is the attic. Typically this is a random wire connected to an antenna tuner. You might have a problem running a lead-in wire from the attic to the room where the receiver is located; the lead-in can be unsightly, kids love to tug on it, dogs like to chew it, and adults have been known to trip over it. One inelegant—but effective—solution is to simply leave the random wire scattered out of sight somewhere in the room where your shortwave radio is located. Possible hiding places include a closet, under furniture (such as a bed or sofa), or even behind the desk, table, or other furniture you place the radio on. Messy? Sure, but it can deliver surprising results when used with an antenna tuner.

Some listeners have had good results with so-called "invisible" antennas. They're not really invisible, but they are harder to notice. A very small wire gauge is used, with #28 and #34 being favorites (these are sometimes known as "magnet" wire). Insulators are hand-constructed from clear plastic tubing or strips. If coax is required, special small diameter types such as RG-174/U are used. The entire antenna is light enough to be supported by kite twine or even rubber bands. These antennas are really difficult to notice, even from adjoining homes, unless you know where to look. The problem is that such antennas aren't very durable. High winds, ice, or other bad weather usually bring one down quickly. Birds also have difficulty noticing the fine wire and sometimes collide with it. (In most cases, the antenna suffers more damage than the bird.)

Listeners living in apartments or condominiums, particularly high-rise buildings, have more difficult problems. Often, there are no support structures or access to the ground. On the upper floors, there can be strong and unpredictable winds. Yet such buildings are invariably constructed of steel and concrete, making an outside antenna necessary.

If you're in such a situation, you have my sympathy. I once lived on the thirty-sixth floor of a 42-story condominium. Rigging up *any* sort of permanent outside antenna was impossible. My solution? I used a 30-foot random wire feeding an antenna tuner. Whenever I wanted to listen, I just opened a window next to my receiver, carefully lowered the random wire down the side of the building, and went SWLing. When I finished, I pulled the wire back inside and set it aside until the next time. Its performance wasn't equal to a good permanently-installed random wire, but it did allow easy reception of international broadcasters and some DX. Moreover, it was safe, inexpensive, didn't bother anyone, and allowed me to enjoy SWLing despite a blanket prohibition by the condominium association on any sort of outside antennas.

Listeners in similar circumstances have used this approach with some interesting variations. One SWL uses a fishing reel to raise and lower the antenna wire! Others get acceptable results just by leaving random wire resting outside a window sill rather than lowering it.

The moral of all this is that you shouldn't despair if you live in a "problem" environment. As long as you can get *some* wire outside and feed its output to an antenna tuner, you can enjoy SWLing!

Active Antennas

An *active antenna* is a physically short antenna element (often measuring less than five feet in length) connected to an amplifier stage. The antenna element in some active antennas is designed for mounting outside and is connected to the amplifier unit by coaxial cable. Other active antennas have the antenna element built into the amplifier unit and are designed for use indoors. Several varieties of active antennas have been introduced in recent years, and each is touted as the ultimate solution for SWLs unable to install a full-size random wire antenna. Even some SWLs who live where they could erect an ordinary antenna have opted for an active antenna.

Are active antennas really as effective as their manufacturers claim? It depends upon the situation. They *can* be useful in some cases. But if you have the space to install a properly tuned random wire antenna or vertical, you'll find those will outperform the active antenna in virtually every instance.

An active antenna is subject to the same rules as other antennas. For example, the antenna element itself must be outside any steel or concrete structure to be effective. An indoor active antenna in such situations will be just as effective—or, more accurately, *ineffective*—as any other type of antenna. Moreover, the amplifier unit will amplify atmospheric and local electrical noise as well as the signal.

The most critical part of an active antenna, and the most likely source of trouble, is the amplifier unit. The amplifier unit is, in effect, an extra receiver RF amplifier stage which precedes the actual receiver RF amplifier stage. It is subject to all the problems that the receiver RF amplifier stage is subject to, including overloading and the spurious signals that result. One particular problem with some active antennas is that signals from strong local AM broadcast band stations can cause spurious signals throughout the lower frequency shortwave bands. This is especially frustrating since active antennas are most likely to be used in urban areas where such strong AM band stations are usually found. To combat this, most active antennas include an attenuator or gain control. Unfortunately, reducing the gain of an active antenna enough to prevent overloading often makes it not very useful.

Amplification of the signal presents another problem. Any amplification of a signal, whether by the receiver or active antenna, introduces some noise into the signal. This is no problem as long as long as the signal is amplified more than the noise level introduced. However, it does mean that "passive" antennas such as random wires will give quieter signals than *any* active antennas. And the noise introduced by an active antenna may vary from frequency to frequency.

Another thing to remember is that an amplifier can't amplify what it doesn't have. There is only so much signal that a short antenna element can absorb and deliver to the amplifier unit, and that signal can be masked by amplifier and natural atmospheric noise. Even a high-performance amplifier unit can't overcome the limitations inherent when using a short antenna.

The performance of active antennas varies widely among different models. A few years ago, an active antenna manufactured by a well-known Japanese company was found in testing to have widely varying gain and noise at different frequencies. At some frequencies, it delivered no gain at all and, coupled with the amplifier noise, produced a signal *worse* than if the amplifier wasn't used! Performance claims for active antennas, such as "performance equal to a 100-foot random wire," should be taken with large grains of salt. Does such a claim refer to a tuned or untuned random wire? Is this true at all frequencies or at just a selected one? Without such information, you can't properly evaluate such claims.

If you decide to try an active antenna, purchase it from a dealer or manufacturer who will allow you to return it after a trial period. Some SWLs swear by active antennas, while others swear at them.

Preamplifiers and Preselectors

A *preamplifier* or *active preselector* is an external RF amplifier stage placed between an antenna and a shortwave receiver. Many also include an antenna tuning circuit in addition to an amplifier section; the amplifier section can usually be switched off and the antenna tuning unit can be used without the signal being given extra amplification.

Many of the same comments regarding a receiver's RF amplifier section and active antennas also apply to preamplifiers and preselectors. Overloading can take place from strong signals on shortwave and the AM broadcast band, although most preamplifiers are better in this regard than active antennas. Gain can be controlled continuously, and the units can be "peaked" for best performance in a desired frequency range.

Preamplifiers and preselectors must be used carefully, since many receiver RF amplifier sections can't handle the added output from them. This is especially so with less expensive receivers; their RF amplifier sections get "swamped" and you have wall-to-wall spurious signals. Excellent dynamic range, as found in quality receivers, is a must. Since the added gain from a preamplifier or preselector is often not necessary, many SWLs use theirs primarily as an antenna tuner and switch in the added gain only when they find a signal too weak to read.

Preamplifiers and preselectors work well with a wide variety of antennas. An outdoor random wire or other antenna connected to a preamplifier or preselector is an ideal arrangement for both SWLing and DXing. Moreover, short antennas can be used with them to make your own "active antenna," usually at a lower cost than an active antenna system.

Antennas for the Broadcast Band and Longwave

Listeners interested in tuning the AM broadcast band and longwave have special antenna needs. The sheer physical size of resonant antennas at those frequencies is a problem; a dipole for 1000 kHz (near the middle of the AM broadcast band) would be 468 feet long! Moreover, many antenna tuners, preselectors, and preamplifiers only operate on frequencies above 1600 kHz. Because of such factors, specialized antenna systems have been developed for use below 1600 kHz.

The most effective antenna for such frequencies is known as the *Beverage*. This is an extremely long (over 1000 feet) wire, in a straight line, installed only a few feet above the ground and with one end connected to ground. The Beverage is very directional, having a reception pattern in the direction of its grounded end which is very narrow. As you might expect, the gain of such an antenna is enormous. While impractical for most SWLs, the performance of Beverages is amazing. Listeners along the east coast of the United States have used Beverages for reception of Australian stations on the AM band!

Far more practical are the so-called loop antennas. Loop antennas come in different types and sizes, but all are compact, indoor antennas which have directional characteristics making it possible to reduce or eliminate interfering stations. A loop has a "figure-8" reception pattern, meaning it picks up little signal from stations located at right angles to its vertical plane. For example, if you are trying to receive a station located to the north or south of you, the loop will reduce the strength of stations on the same frequency located to your east and west. This is known as nulling.

Loops come in two basic varieties. The air core loop consists of a frame shaped like the letter "X" around which several turns of wire are wound. The frame is made of nonmetallic material and is from two to four feet on each side. A ferrite core loop is much smaller, and is a bar or rod of a magnetic material known as ferrite around which several turns of wire have been wound. A ferrite loop is enclosed inside a plastic or metal housing (the latter with openings at each end to provide a signal path to the antenna). Both types of loops can rotate, and most allow the antenna to be tilted as well. Since loops are less efficient than wire antennas, amplifier stages for the AM broadcast band and longwave frequencies are normally used. These amplifiers are built into the loop structure and are used to tune the loop as well. The noise introduced by the amplifier is not as serious on the AM broadcast band, since atmospheric noise is already high on such frequencies.

The output impedance of most loops is a few hundred ohms

and they must be connected to the high impedance input of a receiver.

Some loops are also becoming available for the lower shortwave bands. These usually aren't as useful as loops for longwave or the AM broadcast band, since it's more difficult to "null" a station with a loop on frequencies above 1600 kHz. However, there are some situations where they can be helpful.

If you're seriously interested in DXing the standard broadcast band or longwave, a loop antenna is a necessity. If you have some construction skills, you can build one for yourself using plans available from clubs specializing in AM broadcast DXing. Assembled and working units are sold by SWL equipment suppliers.

Audio Filters

An *audio filter* is a device which processes the audio output of your receiver, allowing audio frequencies in certain ranges to pass unhindered to your speaker or headphones while attenuating or blocking other audio frequencies. Audio filters are often poorly understood, and are the subject of considerable hyperbole by some manufacturers. (One claims its audio filter can improve a radio's selectivity by up to 200 to 1!) The best way to appreciate the usefulness of audio filters is to think of them for what they really are—wide range tone controls.

No audio filter can make up for poor selectivity in a receiver's IF section. If a strong signal from a nearby frequency is overpowering a desired one in the receiver's IF stages, then no audio filter will enable you to hear the desired signal. The specifications for an audio filter, such as "2 kHz selectivity at 6 dB down," refer to *audio* frequencies rather than radio frequencies. What an audio filter can do, however, is clean up the audio output of a receiver and make the final signal you hear more intelligible.

Audio filters are capable of four distinct functions. One is the high-pass function, which means the filter will allow all audio frequencies above a desired one to pass with attenuation while sharply reducing or eliminating lower frequencies. A low-pass function does the opposite, passing all frequencies below a desired one and attenuating the rest. A peak function allows a certain range of frequencies to pass without attenuation but rejecting all above and below that range. Finally, a notch function works like the IF notch filters discussed in the last chapter but at audio frequencies.

Audio filters are generally most effective when used for reception of SSB and CW signals, which are "low fidelity" to begin with. The high-pass and low-pass functions are useful in reducing noise, hum, and "crud" present in a receiver's audio output. The peak function is especially effective for CW, since a much narrower bandwidth can be used for CW (which consists only of a single tone). The peak function is less useful with complex audio of music and the human voice, but you can use it to tailor the audio output to your hearing peculiarities.

Audio filters can be a useful complement to your receiver, even if they can't replace good IF selectivity. A receiver with excellent IF selectivity used in conjunction with an audio filter is a great combination for difficult receiving situations.

Outboard AM Reception Enhancers

Most modern shortwave receivers, especially more expensive communications sets, are designed primarily with SSB reception in mind. AM reception is almost an afterthought on these sets. Audio quality in the AM mode is often poor, with poor audio fidelity, and there are often limited selectivity options in the AM mode. As a result, many experienced SWLs prefer to use older receivers from the 1950s and 1960s; these lack features such as direct frequency readout but do have better performance in the AM mode.

In recent years, some smaller manufacturers have attempted to solve these problems by introducing *outboard AM reception enhancers*. These are complete IF amplifiers and AM detectors which take the receiver's IF signal (usually at 455 kHz) and process AM signals independently of the rest of the receiver. The result is better audio quality and more selectivity features in the AM mode.

An example of such units is the MAP multiband AM pickup from Kiwa Electronics. This unit takes the 455 kHz IF signal from a receiver and provides external IF filtering, synchronous detection, and audio filtering. The audio amplifier section is optimized for best fidelity, and drives an internal speaker. The MAP has a choice of wide and narrow AM filters with shape factors of approximately 1.5:1, an audio notch filter tunable from 1.1 to 4.5 kHz, and a wideband "tone tilt" control to adjust the audio response to your preference. The MAP is connected to a receiver using a "microclip" or inductive probe; no modification to the receiver or soldering is necessary.

Users of the MAP and similar units have found they can dramatically improve the quality of AM shortwave broadcasts. Units such as the MAP are useful for DXers as well as SWLs, since the poor audio quality of AM reception in many modern shortwave radios can make it difficult to understand weak signals. The improved audio from a unit such as the MAP can mean the difference in whether you're able to identify a faint AM signal. Until shortwave radio manufacturers finally make synchronous detection and quality audio output standard features in their products, units such as the MAP will be useful to serious SWLs and DXers.

Radioteletype Receiving Equipment

Perhaps the biggest recent development in SWLing has been the introduction of easy-to-use RTTY receiving equipment. In the past, RTTY receiving gear was military or commercial surplus and largely mechanical; it took up a lot of room, made a lot of noise, and wasn't very reliable or versatile. Today's RTTY

FIGURE 4-4

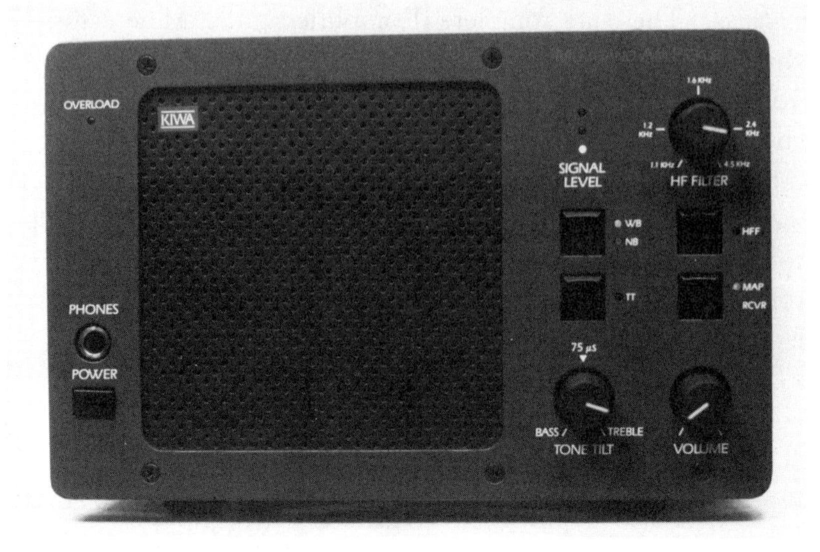

Outboard AM reception enhancers such as the MAP from Kiwa Electronics can greatly improve reception of AM signals on most receivers.

gear is based upon microcomputer technology and has resulted in many SWLs becoming interested in monitoring the "tweedle" stations.

So-called *dedicated* or *multimode* RTTY terminals take audio output from a receiver and convert it into a form which can be displayed on a video screen or printed out on a printer. The new generation of terminals can accept and process a wide variety of signals, such as CW and the ASCII code used with personal computers. (Ham radio operators and some others use ASCII.) These terminals are extraordinarily versatile; they can handle a wide range of RTTY and CW speeds as well as RTTY frequency shifts. Some can also process facsimile and SSTV signals or even RTTY using the Cyrillic alphabet. An external video monitor and printer are usually necessary with these units.

A personal computer interface is used in conjunction with

The M-900 from Universal Radio is one of the new generation of multimode RTTY terminals.

appropriate software. Software is available for most popular personal computers; the job of the interface is to take the output of the receiver and convert it into a form that can be used by the personal computer and software.

With more SWLs having a personal computer available, the personal computer interface approach is usually how most start monitoring RTTY. If your main interests are simply monitoring (or, more accurately, seeing) international press dispatches or ham RTTY, a personal computer interface is a good choice. If you get serious about RTTY and related modes, however, a multimode terminal system is best.

Headphones

You'll soon discover that many interesting signals and rare DX are only heard during odd hours, such as 3:00 a.m. your local time. If you share your living quarters with a family or roommates, you'll find that a good pair of headphones is the only thing to prevent you becoming an outcast! An additional advantage of headphones is that they let you concentrate better on weak and noisy signals by shutting out distracting sounds.

The best headphones for SWLing aren't the kind you use with a stereo system. Those headphones are designed for the broadest possible frequency response. That's great for listening to music, but on shortwave it means you'll also hear static and noise clearly! Instead, *communications* headphones are designed with an emphasis on being able to clearly understand the audio rather than reproducing it with maximum fidelity. Communications headphones are available from SWL equipment dealers, and many receiver manufacturers offer matching headphones for their sets.

CHAPTER FIVE

Radio Propagation

PROPAGATION IS THE TERM used to describe the process by which a radio (or television) signal gets from the station's transmitter to your receiver. There are a lot of different ways a radio signal can get from a distant country to you, and understanding how the propagation works is probably the most important thing for a SWL or DXer to learn. After all, it doesn't matter how good your radio, antenna, and accessories are if propagation conditions won't allow a signal from a desired station to reach you!

Reception from local AM, FM, and TV stations is via *ground wave* or *space wave*. Ground wave (sometimes called surface wave) propagation is best on frequencies below 2000 kHz and is how local AM broadcast band stations are received. As its name implies, the ground wave travels outward from the transmitting antenna along the ground and weakens the further it travels. The amount the ground wave weakens depends upon the type of soil it travels over, with rocky soil absorbing more signal than other types. The least absorption takes place over salt water, and some amazing distances can be spanned by ground waves traveling over ocean paths. AM band DXers along the east coast of the United States are familiar with this phenomenon; listeners located on the Outer Banks of North Carolina can listen to Florida AM stations at local noon, while at the same time listeners on Cape Cod can hear stations from the North Carolina coast. Regardless of the type of surface the ground wave travels over, it weakens as it travels away from the transmitting antenna until it is too weak to be heard above atmospheric noise. Within the area over which the ground wave travels, however, it provides steady, fade-free reception of the station.

As the frequency of a signal increases, the ground wave diminishes. At the frequencies at which TV and FM stations are found (above 54 MHz), the ground wave, for all practical purposes, no longer exists. Local reception at such frequencies is possible because of the space wave (sometimes called the *direct* wave). The space wave travels through open space, much like a beam of light, from the station's antenna to the listener. The range of the space wave is everything that can be seen out to the horizon from the transmitting antenna. If the horizon, as viewed from the antenna, is 100 miles distant, then the space wave can be received over that distance. Actually, the space wave can be received somewhat beyond the actual horizon, since some of the space wave will be diffracted beyond the actual horizon. Space wave propagation explains why FM and TV stations try to use tall transmitting towers or antenna locations atop mountains and skyscrapers. As a general rule, reception via space wave is possible whenever the transmitting and receiving antennas can "see" each other.

Long-distance reception beyond space-wave range on frequencies from 2000 to 30000 kHz (and, to a large extent, on frequencies above 30 MHz) depends upon the Earth's ionosphere. The ionosphere is the part of the atmosphere which extends from approximately 30 to 600 miles above the Earth's surface. The ionosphere can *refract* back to earth radio waves which strike it, allowing reception of the radio waves at a considerable distance from the transmitter. A signal which is refracted off the ionosphere is called the *sky wave*. Figure 5-1 is a simplified illustration of this process.

Refraction off the ionosphere is possible because of ultraviolet light radiated from the sun. The ultraviolet light *ionizes* the

FIGURE 5-1

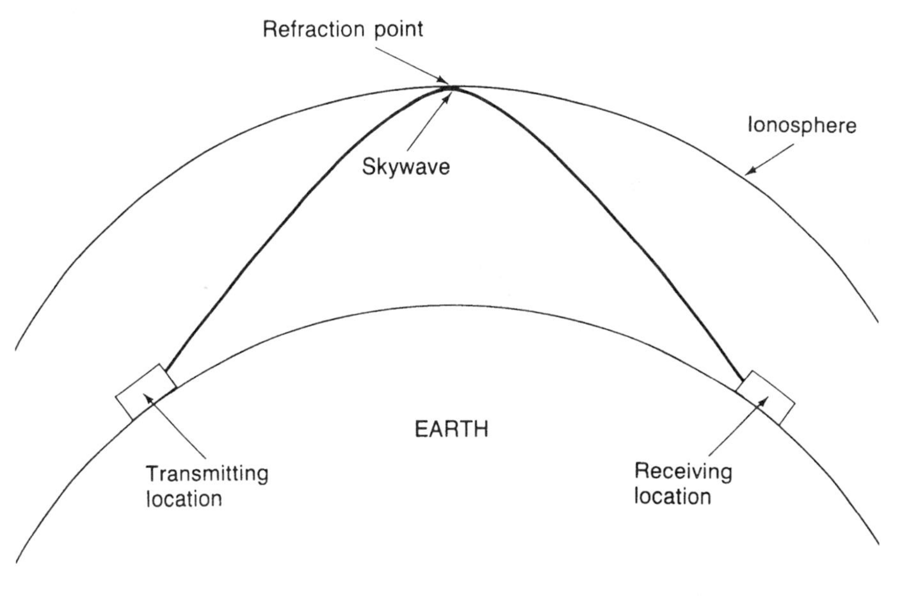

Sky wave propagation via refraction off the Earth's ionosphere.

ionosphere (that is, the atoms of the gases constituting the ionosphere gain an extra electron), turning the ionosphere into a sort of "mirror" capable of refracting radio signals back to Earth. As you might suspect, the ionosphere's ability (or inability) to refract signals depends on the level and type of solar activity.

The Nature of the Ionosphere

The ionosphere is not a smooth, homogeneous layer of the atmosphere. Instead, it's *layered*, and those layers aren't constant. The ionization of the various layers and even their height above the Earth's surface varies with the time of day and season of the year. Nor is the ionosphere constant all around the world. The variation between the hemisphere experiencing summer and the other experiencing winter is especially great, and often a radio signal that originates in "summer" is received in "winter"! The ionosphere is like the world's oceans—they're all part of the same system and are interlinked, but they vary greatly in their characteristics at different locations.

The ionosphere doesn't treat all radio signals the same. Depending on the frequency, the ionosphere at any given time might *absorb* a radio signal instead of refracting it back to Earth. Signals of another frequency would be refracted, but could be refracted by more than one layer. And radio signals of other frequencies might pass through the ionosphere into outer space without any refraction at all. The way the ionosphere "treats" a signal of a given frequency varies. Sometimes the ionosphere is capable of refracting signals of over 50 MHz, while at other times even signals as low as 10 MHz pass through it. Fortunately, the ionosphere is not as unpredictable as it might seem! There are some general patterns it follows, like seasons of the year, and it's not difficult to understand them.

The layer of the ionosphere closest to the Earth is the *D-layer.* It begins at approximately 30 miles above sea level and extends upward to a little over 60 miles. Ionization in the D-layer is the lowest of any part of the ionosphere. Indeed, it's so weak that most radio signals zip through it to the upper reaches of the ionosphere. However, the D-layer usually absorbs some energy from signals passing through it and weakens them. This effect is especially noticeable on medium and high frequency signals. Usually only signals below 300 kHz can be refracted by the D-layer. The ionization of the D-layer is greatest around local noon and quickly drops as sunset approaches. The D-layer, for all practical purposes, ceases to exist at night and during short midwinter days.

The next ionospheric layer is the *E-layer,* which does play a significant role in long-distance propagation. The E-layer is found from approximately 60 to 100 miles above the Earth, although the exact altitude varies from season to season. Like the D-layer, maximum ionization occurs at local noon and decreases as sunset approaches. Unlike the D-layer, the E-layer may retain enough ionization at night to affect propagation. The most curious aspect of the E-layer is the phenomenon known as *sporadic-E* propagation. Cloud-like patches of very intense ionization can form in the E-layer and refract signals, such as those at TV and the FM broadcast band, which normally would pass through all layers of the ionosphere and into space. Propagation of such signals out to approximately 1200 miles is common, and refraction off several different sporadic-E patches (known as *multihop*) over distances of 2500 miles has been observed. In North America, sporadic-E is most common during late May to early August, with some activity reported around the winter solstice. Sporadic-E usually occurs in the morning and early afternoon, although it can show up at any time of the day or year. Since the main impact of sporadic-E is on frequencies above 30 MHz, we'll talk about it more in the section on TV and FM DXing.

The important thing to remember about the E-layer now is that it is especially strong during the day, and it is more absorbent of lower frequency signals. The E-layer is the main reason why you can't hear as many distant signals in the 540 to 1600 kHz AM broadcast band in the day as you can at night; the E-layer blocks them from reaching you. At night, the weakened E-layer is unable to prevent them from reaching the upper layers of the ionosphere and being refracted over long distances.

The *F-layer* is the one you'll be most concerned with, since this is where most long-distance radio propagation takes place on frequencies from 2 to 30 MHz. The F-layer begins at approximately 100 miles above sea level and extends upwards to about 250 miles. During the daytime, the F-layer "splits" into two separate regions known as the *F1*, which extends from 100 to 150 miles in altitude, and the *F2*, which extends upward from the end of the F1. The F1-layer exists in daytime and affects some radio signals, although most signals that can penetrate the

E-layer do the same with the F1-layer. At night, the F1-layer weakens and merges with the F2-layer to form a single F-layer. The F-layer is the last refracting layer; if a radio signal can penetrate it, the signal travels out into space and is lost. The highest frequency the F-layer can refract varies with its ionization. Normally, the maximum frequency is between 20 to 30 MHz but can be lower. However, during exceptional conditions the F-layer can refract signals in excess of 50 MHz.

Since the ionosphere is ionized by the sun, you really can't understand how radio propagation works until you know something about how our sun behaves. The sun follows patterns in its activity, meaning you can observe, measure, and predict its activity, and this allows you to have a good idea of what radio reception conditions on different frequencies will be like. Let's take a look at the most important star in the universe. (At least it is if you're a human!)

The Sun and Radio Reception

The sun's effects on radio reception can be predicted with reasonable accuracy because of four factors:

- the time of day
- the season of the year
- your listening location
- the level of sunspot activity

We can do this because the sun is a relatively stable star compared to many others in the universe (a good thing for life as we know it as well as radio propagation!). But we can't predict solar activity with total accuracy, since the sun is subject to many short-term, erratic phenomena such as solar flares. These unpredictable events do sometimes totally disrupt normal shortwave reception. On the other hand, they may also allow reception of DX stations that otherwise would be impossible to hear.

The ionization of the ionosphere above your head is normally lowest just prior to your local sunrise. As the sun climbs the sky, the ionization increases. At your local *solar* noon (when the sun reaches the highest point in the sky on that day, not when the clock says it's noon), the ionization is at its highest. As the sun moves westward, the level of ionization begins a slow decline. After sunset, the level of ionization begins to fall rapidly and continues throughout the night until sunrise the next morning, when the process starts over again.

If you think the ionization would be greater in the summer than the winter, you're right. Not only are the days longer, but the sun's rays are more direct. (And since the seasons are reversed in the northern and southern hemispheres, the ionization effects are also reversed.) But the result might be different than you expect. During the winter, ionization is usually so low that the F1-layer merges with the F2-layer during the daytime. This produces a single F-layer that is heavily ionized in the daytime but rapidly loses ionization at night. In summer, the more direct angle of the sun's rays produces more ionization of the F-layer. But the ionosphere is warmed more by the sun in summer, and the F-layer splits into its two sections and expands to its greatest height above the Earth. Although the ionization is greater, it's also extended over a larger area, and the resulting ionization density of the F-layer during the daytime is actually *less* in summer than in winter! After sunset, though, the F-layer retains more ionization in summer than in winter. In summer, the F-layers cool and contract into a single, densely ionized F-layer.

Location also affects the ionization of the atmosphere. The sun's rays are more direct and intense at the equator, and as a result the ionosphere is usually most highly ionized around the equator. Ionization is also intense between the Tropics of Capricorn and Cancer. Ionization decreases as you get closer to the polar regions. The ionosphere over Miami will normally be more highly ionized than over Montreal if measurements are taken at the same time. At the poles themselves, some very unusual propagational conditions take place; we'll discuss these later.

Perhaps the most important factor affecting ionization is the level of sunspot activity. The sun emits more ultraviolet radiation when it has more sunspots, and the ionosphere is usually more highly ionized.

Sunspots appear on the sun in a series of cycles. A *sunspot cycle* is defined as the period from a sunspot minimum to a peak number and then down to a minimum again. Cycles take several years to complete, with 11 years being an average cycle. However, sunspot cycles have been observed lasting as few as five years to over 17 years. Reliable records of sunspot activity and cycles have been kept since the mid-eighteenth century. At maximum, the sunspot count of some cycles has exceeded 150 (a sunspot cycle which peaked in March, 1958 had a maximum count of over 200) while the count during a cycle minimum usually drops below 10. The average maximum tends to be around 110.

The number of sunspots changes gradually. Other solar changes are more abrupt and unpredictable. One that causes drastic effects on radio propagation is the *solar flare*. A solar flare is a sudden eruption of gas from the sun's surface which ejects large amounts of ultraviolet light, cosmic radiation, and X-rays. These travel to Earth at the speed of light and start to affect the ionosphere as soon as the flare can be visually detected. Solar flares also release large quantities of charged particles which travel to Earth more slowly, eventually arriving hours to over a day after the flare. One consequence of a solar flare is often a *sudden ionospheric disturbance* (SID). A SID begins as soon as the flare is visible on the sun and can last from a few minutes to a few days. A SID can produce anything from minor disturbances in propagation to complete "blackouts" in which no sky-wave propagation is possible on SW frequencies.

SIDs usually affect only those areas of the ionosphere which are in daylight when the solar flare takes place.

Solar flares and SIDs are the negative flip side of sunspot cycle peaks. Although propagation conditions are usually better during such peaks, the number of solar flares and SIDs increase with the sunspots, and the better propagation conditions are frequently interrupted because of flares.

Another effect of solar flares is the *ionospheric storm*. This is caused by the charged particles that arrive after the flare. The charged particles are attracted by the Earth's magnetic field and enter the ionosphere at the North and South Poles. These charged particles are responsible for auroral displays in areas near the poles, and a brilliant auroral display is almost always accompanied by a disruption in radio communications. However, the effects of "auroral propagation" are noticeable in areas, such as the southern United States, where visual auroras are once-in-a-lifetime events.

The effects of an ionospheric storm are varied and unpredictable. The F-layer sometimes splits into multiple layers or even "disappears" as far as propagation is concerned. The most common effect is increased ionization of the ionosphere and greater absorption of the signals it does refract. However, ionospheric storms of moderate intensity can actually be helpful to DXers. Since charged particles enter at the polar regions, those areas are affected first. The charged particles then work their way toward the equator, but most go no further than the middle latitudes. The ionosphere over tropical regions is seldom affected by ionospheric storms. The result for North American SWLs is that signals from the east or west of their locations may be severely weakened or totally missing, while signals from the south may be unaffected. This means a band normally crowded with powerful international broadcasters from Europe, such as 49-meters, may be free of those stations or the signals may be significantly weaker. This allows reception of rarer domestic broadcasters in South America, Africa, and the Pacific whose signals are generally blocked by the European powerhouses.

Unlike SIDs, the effects of an ionospheric storm are present on both the day and night sides of the Earth. Higher frequencies are usually affected first, with lower frequencies experiencing disturbed conditions later and usually not as severely. (However, the effects of an ionospheric storm can be noticed on the AM broadcast band.) Unlike a SID, conditions during an ionospheric storm gradually deteriorate and return to normal. And ionospheric storms can take place without a SID taking place first. Many flares are not strong enough to cause a SID, but do release substantial quantities of charged particles which do cause ionospheric storms. It won't take too many hours of tuning SW before you'll be able to quickly recognize when conditions are "auroral," usually by the weaker, more "watery" sound of stations to your east or west which are normally loud and reliable.

Propagation Paths

Many things can happen to a radio signal in the interval (measured in millionths of a second) between when it leaves a transmitter and when it is received by you. The signal may be refracted by one or more layers of the ionosphere, or some of its energy might be absorbed by other ionospheric layers. While radio signals "normally" (if there is such a thing in radio propagation!) follow the shortest possible route between two points, it is also possible for propagation to occur along a longer path than the shortest one. Some parts of a signal (such as the sidebands of an AM signal) might arrive at a receiving point out of step, or out of *phase*, with the others due to parts of the signal being propagated differently. Although the difference in the arrival times of the different signal parts at a receiving point is measured in microseconds, this can cause problems, as we'll see later in this chapter. The combination of the physical route a radio signal takes and and how it is propagated via the ionosphere makes up the *path* a radio signal takes. A *multipath* signal means one that arrives at a receiving site simultaneously via different paths. Such signals are also typically distorted.

There are several terms used to describe propagation paths. One is *maximum usable frequency* (MUF). This refers to the highest frequency that can be refracted by the ionosphere along a given path. Any frequency less than or equal to the MUF will be refracted by the ionosphere, and signal strength increases as the station frequency approaches the MUF. This is because signal absorption by the ionosphere *drops* as a signal approaches the MUF. But if a signal's frequency is greater than the MUF, the signal travels through the MUF into space. Since the MUF depends upon the ionization of the various layers of the ionosphere, it varies hour to hour. MUF usually increases along paths in daylight and drops along those in night.

The opposite of the MUF is the *lowest usable frequency* (LUF), which is the lowest frequency on which an intelligible signal can be transmitted over a given path. At frequencies below the LUF, the ionosphere will absorb too much of a signal to make the path usable. Unlike MUF, LUF depends to a degree upon transmitter power and receiving equipment used. There are limits, however; there comes a point beyond which no practical (or even theoretical) improvements in transmitter power or receiving equipment will lower LUF. Like MUF, LUF is constantly changing and is lower along nighttime paths and higher on daytime paths.

The concepts of LUF and MUF are another way to reiterate a point we've mentioned before—while technology has certainly made SWLing easier, SW reception still depends on some natural forces beyond our control. If you're trying to hear a station whose operating frequency is above the MUF, nothing will allow you to hear that station. You can try a better receiver, switch to a longer antenna, add all the preamplifiers or audio filters you want—you'll still hear only noise, not the station.

The same thing will happen if you try to hear a station operating below the LUF for a given path. The only way you can have reliable, daily reception of a station is to be within its ground or space wave coverage. Fortunately, daily changes in the MUF and LUF are gradual and largely predictable. It's possible to make reasonably accurate predictions of reception conditions along various paths. And as for unpredictable events such as SIDs and ionospheric storms—well, they keep SWLing, and especially DXing, challenging!

Since the F-layer is generally more densely ionized during daytime in winter than in summer, the daytime path MUF is usually higher in winter than in summer. However, since the total ionization is less, the nighttime path MUF is normally lower in winter. The lower daytime MUF in summer is compensated for by a higher MUF at night. The effect of these seasonal variations is to make certain paths and frequencies useful at some times of the year but not at others. Let's suppose you're an SWL in eastern North America. In winter, frequencies above 15 MHz are often very good during the morning for reception of stations located in eastern Africa and the Indian subcontinent. During summer, those same stations are seldom heard on such frequencies. But summer is not a total loss. During wintertime late evenings and nights, very little will be heard above 15 MHz in eastern North America. In the summer, however, 15 MHz and above is alive during those same hours with signals from Asia, the Pacific, and Australia.

Great Circles, Long Hauls, and Multiple Hops

The *great circle path* is the shortest direct route possible between two points on Earth. This is the route a radio signal will take whenever possible. Whether it is possible depends upon the signal's frequency and the MUF along the great circle path. Sometimes reception is possible along a much longer path between two stations, and this is known as *long haul path* reception. The long path is formally defined as the reciprocal of the great circle path between two points, but is usually used to mean any path lying in the opposite direction from the great circle path. As a general rule, you'll find that great circle paths are located along daylight paths while long haul ones lie in darkness.

If you really want to understand this business of great circle and long haul paths, spend some time with a globe and some string or a rubber band. (Maps and atlases won't work, since any flat representation of the Earth's surface is unavoidably distorted and misleading.) You'll quickly discover that the shortest path between two places on Earth is never a straight line—it's a *curved* line. (All true straight lines, in the geometric sense, between two points on Earth must pass under the Earth's surface.) This is not just a matter of academic interest, because it means that signals will arrive at your location by paths you might not expect. For example, the shortest path between much of North America and Asia and Europe lies over the North Pole. Understanding where the true great circle and long haul paths lie is important because it is propagation along them that determines whether or not you can hear a desired station.

Not every great circle or long haul path can support propagation at a desired frequency or to a given area. Let's suppose it is early afternoon in July, you're in eastern North America, and you want to hear a station from Australia. At that time, however, all of Australia is in darkness and it will be at least a couple of more hours before sunrise in eastern Australia. In this situation, it's very unlikely that any path or frequency can support the necessary propagation. (It's not *impossible*—just very unlikely.) But a few hours later, all of Australia is in daylight and there's usually an excellent path between the two points on frequencies in the 14000 to 18000 kHz range.

This example shows how certain paths are *open* or *closed* at different times of the day or year. Propagation between distant points can get very complicated, with some parts of a path in daylight and others in darkness. Since the Earth rotates, these patterns of daylight and darkness, along with MUF and LUF, are constantly changing. When a path first is able to support propagation at a given frequency, it is said to *open*. When it is no longer able to support propagation, the path is said to *close*. You'll often hear SWLs talk about band *openings*, which refer to those periods when propagation is possible from certain areas on various frequencies. The length of an opening can range from a few minutes to most of the day. It's not uncommon to be able to hear major international broadcasters from Europe for several continuous hours. However, DXers in eastern North America may find the 60- and 90-meter broadcasting bands open to India for as little as ten to fifteen minutes around sunrise near the winter solstice. Yet these brief openings are eagerly sought by such DXers because they offer the only opportunities to hear such stations in eastern North America.

So far we've discussed propagation with the implicit assumption that a signal is refracted only once by the ionosphere. This is sometimes the case, but more often a signal is refracted by the ionosphere, returns to Earth, and then "bounces" back to the ionosphere to be refracted once again. This is called *multiple hop* (or "multihop," the same term used earlier to describe some sporadic-E receptions). Multiple hop reception is common on paths of more than a few thousands of miles, and you'll soon learn to recognize it by rapid, rhythmic variations in signal level. On your radio's S-meter a "long" multiple hop signal will "bounce" the S-meter rapidly.

Multiple hop signals are weaker than those from more direct propagation routes because some energy is absorbed each time a signal is refracted by the ionosphere or returned from Earth. The ground is particularly absorptive of radio energy, but sea water weakens a signal much less. Thus, multiple hop signals which have return points on the oceans are usually stronger than those which involve large land masses, such as Asia. Some of the best DX receiving locations are those where the last few Earth returns involve oceans, such as along coastlines or on islands. Hawaii and New Zealand are those locations where outstanding DX receptions have been reported for many years.

Multiple hop signals tend to be more stable on frequencies close to the MUF for a given path. While such signals vary in strength, they are usually strong enough for complete intelligibility. That's not the case as the frequency drops from the MUF. Lower frequency signals may fade out altogether during multiple hop reception; they might be audible for only one minute during a three-minute reception period. Fading patterns on low frequency signals tend to be slower than on higher frequencies; the rise and fall in signal strength is not as abrupt.

An interesting effect is noticed when signals pass through either polar region. Each pole is surrounded by an *auroral zone*. This is the region in the E-layer where auroral activity takes place, and one surrounds each pole. Even when no aurora is visible, there is usually enough irregular ionization in these regions to disturb radio signals passing through them. This is usually manifested by a rapid, irregular fluctuation of the signal known as *flutter*. Signals can fluctuate in strength at a rate in excess of 100 variations per minute. The fluctuations can be so rapid that voice communications, such as AM and SSB, become unreadable. You'll quickly become familiar with auroral flutter by its effects on signals from Asia arriving over the North Pole.

Combined Propagation Paths

Ionospheric refraction is often a lot more complex than we've assumed so far. Suppose that both the F1-layer and F2-layer are present in the ionosphere; a signal could be (and often is) refracted off both layers. Since the F2-layer is higher than the F1-layer, any signal refracted off it would have to travel further from the transmitter site to the refraction point and back to

Earth. If the same signal is being refracted off the F1-layer, the signal from the F1-layer would arrive at the eventual receiving site *ahead* of the signal refracted off the F2-layer. And, as mentioned earlier, it's possible for the sidebands and other parts of a signal to be propagated by different paths. Sometimes one sideband will be propagated but the other will not—propagation effects can vary with just a few kHz! When these conflicting signals are processed by a receiver, the mixer stage gets "confused" and the result is distorted audio. If the signals are completely out of phase with each other, they wind up cancelling each other out! And the situation gets even more complex if the E-layer gets in the act and is also refracting some of the signal.

A special case also happens at night on the standard AM broadcast band when listening to stations within a few hundred miles of you. The station's signal can arrive via both ground wave and sky wave. The ground wave signal arrives microseconds ahead of the sky wave and produces distorted audio in receivers. The "phase cancellation" effect mentioned in the previous paragraph is especially noticeable on the AM broadcast band.

These multiple propagation paths can change rapidly. The F2-layer might suddenly be unable to support propagation but the F1-layer might still remain open. Or the E-layer could become sufficiently ionized to block propagation. These changes are particularly noticeable if a station is weak or the path difficult. Every experienced SWL has stories of how a weak signal abruptly faded into the noise and was lost or, more happily, a barely audible station suddenly increases in strength to full readability.

The atmosphere is sometimes described as an ocean of gases instead of water, and that analogy is apt for the ionosphere. Like the ocean, it is dynamic and in constant change. The levels of ionization and the layers involved in refracting a signal are always in a state of flux. This is particularly so if the terminator for sunrise or sunset sweeps across any part of a path (all daylight or all darkness paths tend to be more stable). Don't be too disappointed if a path isn't open that "should" be—and don't be too surprised if a path is open when it "shouldn't" be!

Refraction Angles and Skip Zones

You might be puzzled when you can hear stations more distant from you while closer stations—even those using higher transmitter power—aren't audible. This is due to the *skip zone* phenomenon. This is the area beyond a station's ground or space wave coverage but short of the area in which the sky wave returns to Earth. The signal literally "skips over" SWLs located in the skip zone. Figure 5-2 shows how a skip zone operates. Although only one refraction is shown in figure 5-2, skip zones are also found in multiple hop reception. A separate skip zone is found under each refraction point. However, since multiple hop reception often involves several layers of the ionosphere, a station is seldom totally inaudible along a multiple hop path for long.

Skip zones aren't as neatly static as figure 5-2 suggests. Even for the same station on the same frequency, there are usually one or two periods per day when areas that can normally receive the station are themselves in a skip zone. These periods are usually around sunrise and sunset, when reception conditions are changing from night to day and back again. For example, WWV on 15 MHz can normally be heard throughout eastern North America. But shortly after sunset, eastern North America is often in a skip zone for WWV for a few minutes, and WWVH in Hawaii, also on 15 MHz, is sometimes heard atop WWV at that time.

The distance covered by sky wave propagation depends upon the *angle of refraction*, also known as the *critical angle*. The angle is defined in reference to the Earth, and the lower the angle (as measured in degrees), the longer the distance covered by the

FIGURE 5-2

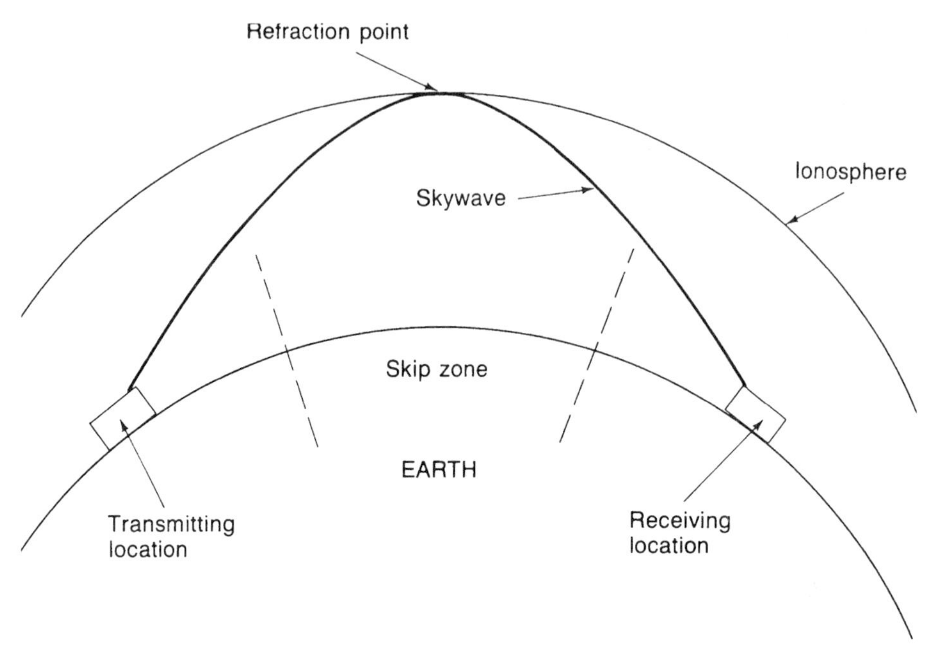

Although located closer to the transmitting station, listeners in the skip zone will be unable to hear the station because the signal is refracted overhead.

sky wave. Higher frequency signals have a lower angle of radiation than low-frequency signals. Figure 5-3 shows how angles of radiation affect the distance covered by a signal. This explains why it is generally more difficult to receive a lower frequency signal than a higher frequency one over the same distance. The lower frequency signal requires more refractions and Earth returns to cover the same distance, with each refraction and Earth return weakening the signal.

The distance covered by a signal also depends upon the altitude at which refraction takes place. A refraction from the F2-layer covers more distance than one off the F1-layer. And the distance covered by a refraction increases as the signal's frequency approaches the MUF. The greatest distance covered by sky wave propagation and the longest skip zone are the

FIGURE 5-3

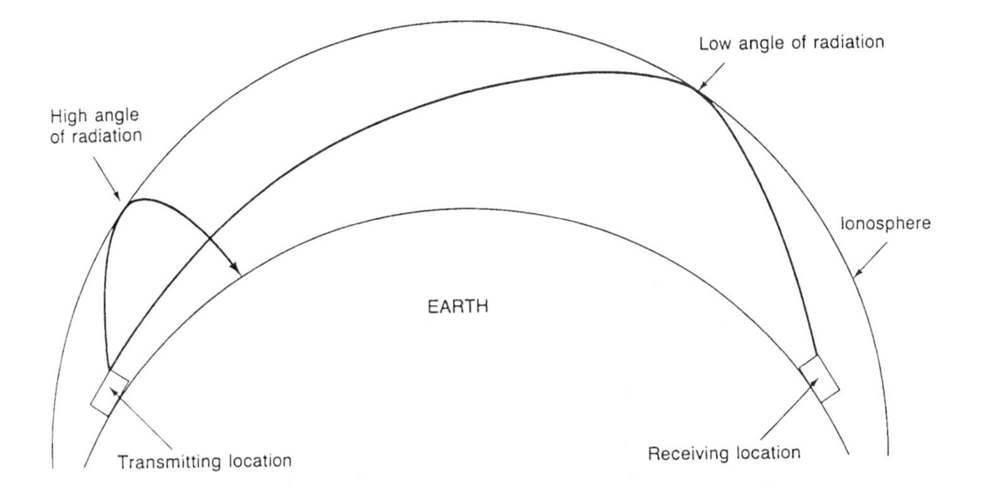

The lower the angle of radiation, the more distant the propagation of a signal.

products of a low angle of radiation coupled with an operating frequency near the MUF. (Low-frequency signals are seldom near the MUF or refracted by the F2-layer, and this is another reason why they are more difficult to receive over long distances.)

Many stations use transmitting antennas designed to concentrate energy into a low angle of radiation. Vertical receiving antennas also give better reception of signals arriving at a low angle. However, the low angle of radiation at which many DX signals arrive can mean problems if you live in a mountainous area. For example, if there are high mountains to your east, you might find reception of weak DX stations from that direction to be difficult. This is because the mountains block and absorb enough energy from such weak signals to make them inaudible. Mountains pose no problem for reception of most shortwave stations, but you probably won't be too surprised to learn that many top DXers have selected homesites on hilltops!

Reception Patterns

You might be puzzled at how some shortwave frequencies are useful for reception at times you wouldn't expect. For example, the 19-meter international broadcasting band often permits reception of stations located to the west until late at night. The 49-meter and 60-meter broadcasting bands are normally thought of as night bands, yet they often let you hear stations located to the east in early afternoon and stations to your west until a couple of hours after local sunrise. The 31-meter and 25-meter bands are often open to some part of the world around the clock. You'll soon notice that there are many such exceptions to the general guidelines given back in chapter 2. Why do these happen?

Let's take a look at what happens along a propagation path. Perhaps the biggest change happens when it's sunrise or sunset at the transmitter site. As sunset approaches, the D-layer and E-layer near a transmitting location rapidly weaken or disappear. This allows signals to reach the F-layer (or layers) without attenuation. Since low-frequency signals are especially affected by these layers, propagation becomes much easier on these frequencies at sunset. If you're listening in North America, it's easy to tell when the sunset terminator has moved across a location in the Pacific or Asia, because stations on low frequencies that were inaudible only a few minutes previously will suddenly and dramatically increase in strength. A similar effect happens at sunrise; the D-layer and E-layer reform within minutes after sunrise and absorption of low frequency signals returns quickly to daytime highs. Listeners in North America can often follow sunrise across Africa after 0500 UTC by listening to various African broadcasters on 60-meters fade into the noise as sunrise moves across the nations.

Sunrise and sunset at your listening site isn't as dramatic, because the key factor is the state of the ionosphere at the various refraction points along the path. This means low-frequency signals from the east can fade in long before sunset at your receiving location, and high-frequency signals from the east can be heard before your sunrise. Conversely, high-frequency signals from the west can be heard well after your sunset and low-frequency signals from the west can be heard after your sunrise.

Advanced DXers take advantage of a technique known as *gray line propagation*. The gray line is the twilight or dawn terminator that separates daylight and darkness areas of the Earth. Gray line propagation takes place whenever a receiving and transmitting site are both in the terminator area. At your listening location, gray line propagation is usually possible a half hour before and after your local sunrise and sunset. The D-layer and E-layer are both virtually absent at both the receiving and transmitting sites during this period, and exceptional DX is possible (such as the openings to India on 60-meters and 90-meters mentioned earlier).

With experience, you will soon become familiar with various reception patterns and what can—and cannot—be heard from your location. Experienced SWLs often have "beacon" stations for various bands to give them a quick indication of reception conditions. Major international broadcasters are usually not good beacons, since their power transmitters and effective antennas can deliver strong signals even in poor reception conditions. Far better are domestic broadcasters or standard time and frequency stations. Most DXers have a list of beacon stations around the world for different times and frequencies. If a beacon station can't be heard, it is unlikely that other stations from that area in the same frequency range can be heard. However, if a beacon station's signal is stronger than usual, then it's a good indication that DX stations from the same area can be heard in that frequency range.

Propagation Forecasting

Most major international broadcasters employ at least one specialist whose responsibility is forecasting propagation conditions along with suitable frequencies and times for paths to the station's intended target audiences. Thus, if your primary SWLing interest is in receiving major international broadcasters, you don't have to worry about the best times and frequencies for reception; those have already been selected for you. However, if you're more interested in DX targets, some simple propagation forecasting techniques will greatly increase your chances of hearing those stations. While it's impossible to predict such events as SIDs with any real accuracy, other solar phenomena are more regular in their behavior.

One key in propagation forecasting is the sun's rotation period, which is approximately 27.5 days. For example, suppose that a large group of sunspots is present on the face of the sun. This indicates high ultraviolet radiation from the sun and heavy ionization in the ionosphere as a result. As the sun rotates, the sunspot group moves away from the Earth's view and the level of ultraviolet radiation also drops. But after 27 days, the sunspot group again rotates so it can be seen from the Earth, and the level of ultraviolet radiation is again high. This means that solar effects on propagation tend to repeat in 27-day or 28-day cycles. The exact propagation conditions seldom repeat, since the sun is a dynamic body and changes can occur during its rotational period. But if you keep records of propagation conditions for a given day, and make appropriate adjustment for seasonal changes such as the amount of daylight and angle of solar radiation, it's possible to predict conditions 27 days in advance with accuracy as high as 80 to 90 percent. Sometimes conditions will be better (the sunspot group gets larger) or worse (the sunspot group shrinks or a solar flare occurs). But you'll find the 27-day solar rotation cycle to be a good general guide to propagation conditions. This is especially true in years of low solar activity and few sunspots, since the sun is more "stable" then.

Sometimes the effects of flares repeat every 27 days. A weak flare which is just beginning when it rotates out of view may not deliver its full impact until it reappears. A flare spotted during a previous rotation can also have a diminished effect when it again rotates into view.

A daily indication of solar activity is provided by radio stations WWV and WWVH at 18 minutes past the hour on each of their operating frequencies. At this time, WWV and WWVH broadcast a prediction of the Earth's geomagnetic activity along with the *solar flux* and *K-index* readings. The geomagnetic activity report is fairly straightforward and uses terms such as "quiet," "unsettled," or "active" to describe conditions; special reports are given when an ionospheric storm is in progress. These reports indicate how active the ionosphere is. An "active" or "unsettled" report means that propagation conditions may be shifting rapidly and that signals may be affected adversely; this is especially true for signals traveling along paths between points in higher latitudes. An ionospheric storm indicates that "auroral conditions" are being experienced. By contrast, a "quiet" report usually means that most propagation paths are not significantly disturbed by ionospheric events.

Even if the paths are not disturbed, this does not indicate what the MUF or LUF along a path might be. One measure of the intensity of solar radiation and resulting ionization is the solar flux measurement. This is a measurement of radio noise "transmitted" by the sun on approximately 2800 MHz and correlates closely with ultraviolet radiation from the sun. The higher the solar flux reading, the higher the MUF will be on most paths.

The K-index is a measure of activity in the Earth's magnetic field as measured at Boulder, Colorado. A major factor influencing the K-index is charged particles from the sun. A small K-index number indicates good propagation between high latitude points and low ionospheric absorption, while a large K-index reading indicates the possibility of auroral conditions and high absorption along paths in higher latitudes.

While WWV/WWVH broadcasts propagation information each hour, it is not updated that regularly. The solar flux value is the previous day's reading; however, solar flux seldom varies significantly in a day. The K-index is updated each three hours and is usually no more than six hours old when given. In fact, it is possible to get the latest information via telephone; the number is (303) 497-3235. (This is a toll call, and collect calls are not accepted.)

How do you interpret the WWV/WWVH information? The best conditions for most receptions would be a forecast for quiet geomagnetic activity with a low K-index number (no greater than 2) and a high solar flux reading (such as 90 or above). Similar conditions but with a low solar flux number means that the MUF may be lower and thus high frequencies may not be as useful (although low-frequency DX may be excellent). A geomagnetic activity report of "unsettled" or "active" but with a low K-index number and high solar flux indicates average or normal reception conditions. If the K-index is high (up to 4) or the solar flux is low, conditions will be well below normal. If the K-index is high (up to 4) or the solar flux is low, conditions will be below normal. If an ionospheric storm is in progress, or if the K-index is over 5, reception conditions will be disturbed.

Standard time and frequency stations in other parts of the world, such as Japan's JJY, also give propagation forecasts, as well as major international broadcasters with programs for SWLs, such as Radio Canada International.

Several tools to assist propagation forecasting are available. One useful example is "The DX Edge," a device similar to an oversized slide rule. It consists of a world map with clear plastic overlays representing the day and night patterns for all twelve months of the year. This allows instant determination of which areas of the world are in daylight or darkness and where the gray line terminator lies. "The DX Edge" is available from several

SWL equipment suppliers or directly from Xantek, Inc., P. O. Box 834, Madison Square Station, New York, NY, 10159. In addition, there are several propagation prediction programs now available for popular makes of personal computers. These programs take input such as the date and WWV/WWVH figures for the solar flux to produce predictions of MUF and LUF along paths between various points.

CHAPTER SIX

Major International Shortwave Broadcasters

IF YOU WERE TO ASK long-time SWLs which was the first station they heard on shortwave, the answer would probably be something like "the BBC," "Radio Nederland," "Radio Moscow," or "HCJB." These are just a few of the well-known international shortwave broadcasters beaming powerful signals to North America each day. It doesn't take any effort to hear such stations; a quick scan of the 49-meter through 19-meter broadcasting bands between 0100 to 0500 UTC will let you hear dozens of different countries broadcasting in English.

The signals from major international broadcasters are the only ones on shortwave that can be legitimately termed "easy listening." Their high power and optimized propagation paths reduce the detrimental effects of the ionosphere. The programs themselves are intended for listeners in other countries, and some attempt is usually made (but not always successfully!) to take into account the interests and preferences of listeners in the target country.

Some SWLs who start off by tuning international broadcasters move on to more specialized interests, such as pure DXing or utility listening, but many stick with listening to the major international stations. (Even hardcore DXers usually have a couple of stations they listen to on a regular basis.) Even though shortwave radio is the oldest electronic channel of international communications, it still has several advantages over more contemporary methods. Perhaps the biggest is that shortwave radio can leapfrog the "gatekeepers" of North American media. These gatekeepers determine what is to appear in a newspaper, magazine, and radio and television newscasts. They serve a necessary function, since only so much news can appear in a single newscast or newspaper, but they also screen out much that might be of interest to some people. Broadcasts from international stations are the only way most North Americans can learn about domestic issues and events in many foreign countries.

But international shortwave broadcasting has its problems. As mentioned in chapter 1, many (if not most) international stations are little more than public relations vehicles for the nations funding them. This is true for stations across the political spectrum; the Voice of America isn't as strident as Radio Beijing, but the purpose of both is to create a favorable image for the operating nation. Moreover, the gatekeeping function is still performed, but this time by employees of the international broadcasters (often with the aid of some government bureaucrats). And a lot of international shortwave programming is dull. Admittedly, "dull" is a subjective state and what one person finds fascinating another finds dull. But you will likely find much programming from international stations to be boring. With many stations, you'll have to sift through a lot of tedious material to find the real nuggets. But, as you'll find in the stations profiles to follow, some truly electrifying moments can be heard on major international broadcasters.

Leaders of emerging nations seem to believe in the ability of an international broadcasting station to generate a favorable image of a nation. It seems as if one of the first things many developing nations do after building a major international airport is to set up an international broadcasting service! Other believers in the power of shortwave radio are evangelical Christian organizations, which operate several shortwave broadcasters around the world. Such stations often reflect little or nothing of the countries in which they operate. For example, HCJB in Quito, Ecuador is often the first station new SWLs hear from Latin America. (It may also be the first station new SWLs hear on SW, as was my case!) However, you won't learn a great deal about the culture, politics, and lifestyles of Ecuador from HCJB, because that's not its prime purpose—evangelism is.

One disconcerting habit of international broadcasters is the way they change frequencies and times of broadcasts at different times of the year. Most stations do so four times per year, usually around the first Sunday in March, May, September, and November. Stations don't change frequencies and schedules just to be perverse; instead, they're trying to take advantage of seasonal propagation changes and expected MUFs. But this does mean you can't rely on always hearing a station on the same frequency the way you can your local AM, FM, and TV stations. It also means this book won't give you frequencies for international broadcasting stations, with the exception of some "traditional" frequencies which certain broadcasters have used year-round for years. To keep up to date with the latest times and frequencies for these stations, consult a reference such as *Passport to World Band Radio* or magazines such as *Popular Communications* and *Monitoring Times*. Membership in a lively SWL club, such as the North American Shortwave Association (NASWA), is also worthwhile. Finally, you can write stations directly and ask to be put on their mailing list for program schedules. This chapter will contain addresses for many of the best-known stations.

International shortwave broadcasting is undergoing great changes as this book was being written. Some nations, such as South Africa, have greatly curtailed their broadcasting activities and others, such as Canada, have considered doing so. On the other hand, the major changes in the Soviet bloc in 1989 and 1990 have been reflected in the international broadcasting services of those nations, with many genuinely dramatic moments being heard by listeners. The years ahead in international broadcasting promise to be equally tumultuous and exciting!

The British Broadcasting Corporation (BBC)

Amid all the change in international broadcasting, there is one "rock"—the British Broadcasting Corporation (BBC). The BBC epitomizes international SW broadcasting for many people. Once you listen to it, you'll understand why. For program quality, news accuracy, and worldwide coverage, the BBC is in a class by itself. While no one at the BBC has come out and said so explicitly, the BBC seems to perceive its programming as a service to all the people of the world. It comes very close to that.

The BBC is an independent public corporation supported primarily by license fees paid by users of radio and television sets in the United Kingdom. It is overseen by a board of governors composed of noted and respected persons in British public life. Day-to-day management is in the hands of career broadcasting professionals. This arrangement seems to work well, since the BBC is perhaps freer of governmental influence than any other publicly financed broadcasting organization. In addition to its shortwave services, the BBC also operates four radio networks on AM, numerous local AM and FM stations, and television networks. There are also regional broadcasting services, such as BBC Radio Scotland, for the other components of the United Kingdom. The service that SWLs are familiar with is known as the BBC World Service and operates on various frequencies 24 hours per day.

For years, the BBC had the reputation of being a stuffy old dowager of international broadcasters and was known as "Auntie BBC" even to its fans. Things have changed greatly since then. Although some faint traces of an Oxbridge mentality can sometimes still be detected, a new generation of broadcasting professionals at the BBC understands their audience includes more than the House of Lords. The result has been more lively, innovative, and experimental programming without resorting to schlock.

Many BBC programs are brief. Don't expect hour-long segments devoted to a single program; often, a BBC program may last as little as five or ten minutes. (This is also the case with many other international broadcasters.) While the time allocated to each program may be short, you'll find that the programs don't waste time or words; they are succinct without being skimpy.

What are you likely to hear on the BBC? One staple is something you seldom hear on North American radio—drama. If you've never listened to people "act" with just their voices, you're in for a treat. You'll hear presentations of classic works such as *Les Miserables* and *Treasure Island* as well as the efforts of more contemporary playwrights and writers. Other frequent presentations are programs devoted to a single major topic or theme, such as dyslexia or hunger. These are different from the "investigative" pieces run by American television; a more accurate term for them would be "research" programs. The BBC emphasizes a thorough digging into the facts of a given situation and a balanced presentation of differing opinions rather than an emotional approach designed to get ratings. The BBC, as perhaps the most independent of all government-financed broadcasters, is not afraid to tackle sensitive subjects even at the risk of offending the party occupying 10 Downing Street. Many of these programs deal with domestic British issues that might not have much relevance for overseas listeners, although the BBC does examine many issues of global importance.

Perhaps the best-known feature of the BBC is its news broadcasts. While objectivity in news presentation is impossible to measure, the BBC's efforts at being impartial, fair, and balanced are recognized worldwide. In fact, the BBC may well be the most trusted news source in the world today. It is not uncommon in many parts of the world for listeners to tune to the BBC to find out what is happening in their own country. Many Americans traveling abroad turn to the BBC instead of the Voice of America to find out what is going on in the world. You'll find BBC news summaries on the hour. These are supplemented frequently by new analysis features and summaries of the British press.

Music is another popular subject for BBC programs. In past years, the BBC stuck to opera, classical, and symphonic works with only passing acknowledgment of more contemporary trends in popular music. In the early 1960s, one would never guess by listening to the BBC that Britain was the epicenter of a revolution in popular music; the "Beeb" still programmed such familiar standbys as Victor Sylvester and his orchestra. There is still much classical programming, but the BBC today also features contemporary British rock stars, jazz, and features on artists of worldwide popularity (such as Frank Sinatra). BBC music programs consist of more than a disc jockey playing records; frequently, the artists themselves are present to talk about their work.

One indication of how much the BBC has changed over the years is how John Peel and David Lee Travers, two well-known disc jockeys for offshore "pirate" radio stations during the mid-1960s, now have rock music programs on the BBC.

The BBC's features, ranging from poetry to features for merchant seaman, are as varied as its audience. There are serials, often with a science fiction theme. And such futuristic fare might be followed by a mundane farming report.

One program that is a particular favorite of mine is "Letter from America," hosted by Alistair Cooke. Cooke has an uncanny ability to come up with insights on American politics, culture, and events that are amusing, entertaining, and thought-provoking.

While only English-language programs have been mentioned so far, the BBC transmits in other languages such as Arabic,

Indonesian, and Urdu. You won't find these programs of much interest unless you understand the language, of course, but they're an important part of the BBC's reputation and following worldwide. Those whose interests lean more toward DXing than SWLing find it valuable to be able to quickly recognize the BBC in other languages to prevent spending time and effort trying to identify a "rare" station that turns out to be the BBC!

In addition to transmitter sites in Britain itself, the BBC also operates relay stations at such locations as Ascension Island, Hong Kong, Cyprus, Oman, and Singapore. It also shares a relay station on the Caribbean island of Antigua with the German broadcaster Deutsche Welle. In addition, the BBC is relayed via Voice of America transmitter sites at Delano, California and Greenville, North Carolina, and also by Radio Canada International from its Sackville, New Brunswick station. (In turn, the BBC makes its British transmitter sites available to those two broadcasters.) However, unless you have a copy of the latest BBC transmission schedule or catch the site location when the station signs on, you won't be able to tell which transmitter site you are listening to, since all identifications are given as "BBC World Service" or "This is London" once a transmission is in progress.

Like most international broadcasters, the BBC uses an *interval signal* to identify itself before programming actually begins. An interval signal is a distinctive sound or musical piece repeated before a station signs on; this helps SWLs find and identify the station's signal. The BBC uses the musical notes "B-B-C" repeated on a tonal scale. You can also hear the bells of London's famous Big Ben between several BBC programs.

The result of such efforts is powerful, reliable signals from the BBC. A quick scan across any international broadcasting band that's open will almost certainly produce a frequency or two in use by the BBC World Service. Like other international broadcasters, the BBC changes the frequencies it uses several times per year. However, there are a few frequencies which the BBC has used without interruption for decades, such as 9410, 12040, and 15070 kHz; the latter is a good choice for daytime reception in North America.

The latest BBC transmission schedule is available from the BBC World Service Publicity, P. O. Box 76, Bush House, Strand, London WC2B 4PH, Great Britain. If you become a fan of the BBC, you might want to subscribe to a monthly program journal known as *London Calling*. This is a bit like *TV Guide*, listing programs by time and day along with descriptions of each program. A sample copy of *London Calling* and subscription information can be obtained when you request a program schedule.

The BBC is flooded with reception reports, and as a result sends out a nonspecific "thank you for writing" card in reply. However, the BBC apparently reads all listener mail, and especially values comments and suggestions concerning their programs. Reports and letters can be sent to the BBC World Service at the same address as for *London Calling*. By the way, the BBC operates a "World Information Centre and Shop" at Bush House; if you're ever in London, you can stop by and obtain BBC publications and souvenirs there.

If it seems like I'm a big fan of the BBC, you're right. My love of the BBC was made permanent in 1986 when I was on a tour of the Soviet Union. I brought along a small portable shortwave radio. On April 26, I arrived in Kiev, unaware (along with the two million citizens of Kiev) that a major accident had occurred at the nearby Chernobyl nuclear power station less than 48 hours earlier. As rumors swept the city, I turned to my SW radio for information. The Voice of America carried the early, sensationalized reports of a nuclear meltdown at Chernobyl. The BBC stuck strictly to what was accurately known about the situation and gave a carefully balanced assessment of the possibilities in situations where the facts were unclear. The Voice of America was frankly alarmist; the BBC, while not downplaying the seriousness of the situation, reported what it knew and avoided guessing to fill in the blanks. In retrospect, the BBC's reliability was proven greater than that of the Voice of America. I'm one American who wishes the Voice of America could do as well as the BBC!

Radio Moscow

In terms of sheer size, Radio Moscow is probably the largest shortwave broadcasting organization in the world. It's almost impossible *not* to be able to hear Radio Moscow in English in North America at any hour of the day or night! If you want to hear Radio Moscow, no special effort is needed; just tune around the major international broadcasting bands during your evening hours and you'll find it on several frequencies. And, unlike years past, you'll probably find something worth listening to. At the time this book was being written, Radio Moscow—and indeed Soviet international broadcasting as a whole—was a mirror of the major forces shaking the Soviet Union to its foundations. Who a few years ago would have imagined that Radio Moscow would be playing David Bowie, running commercials for Soviet products, and advising listeners to "party hearty" at the end of a program called *Vasily's Weekend?* The question many SWLs have is whether this is a permanent shift in Radio Moscow or, as was the case with Radio Beijing, is only a "false spring."

Radio Moscow runs two English language services. The Radio Moscow World Service is patterned after the BBC World Service. Although it's not as interesting or objective as the BBC, the Radio Moscow World Service is even more inescapable; they often use two or more frequencies in the same broadcasting band (at the time this book was being written, the Radio Moscow World Service used five different frequencies in the 31-meter band from 1900 to 2000 UTC!). Radio Moscow also operates a North American Service during evening hours in North America. This service goes out in separate west coast and east coast editions, each timed for best reception in the respective target areas.

The World Service carries programs that are of more "universal" interest. The previously mentioned *Vasily's Weekend* is their version of "American Bandstand," complete with a chattery American-sounding disc jockey, rock and pop music, and even telephone call-ins. Another program called *New Market* describes Soviet products, gives addresses and telephone numbers for more information, and tries to encourage Western investment in the USSR. (The style of presentation on this program is faintly reminiscent of those late-night ads on UHF-TV offering bottle cutters or the collected recordings of Slim Whitman.) And there are newscasts, commentaries, and music (including Soviet jazz on the *Jazz Show*) scattered throughout the hour.

The North American Service also has news, commentaries, and music. Some of its features include the venerable *Moscow Mailbag*, featuring the now legendary Joe Adamov. Joe has been a part of this show for over three decades and, although his role has diminished recently, he still has a certain touch. His forte is answering questions, which often are hostile, in a smooth, polished manner. While the content of his answers has always depended upon what was politically correct at the moment, his style is strictly his own. Sometimes he is serious, while at other times he seems genuinely offended or amused that a question was even asked. With some questions, you can almost picture Joe shaking his head in disbelief that a listener could be so ill-informed. One may disagree sharply with his politics, but most listeners have to admit that Adamov has been an effective spokesman for Soviet interests through the years.

The current superstar of the Soviet media is Vladimir Posner. You can hear him on the North American Service on *Top Priority*, a panel discussion on domestic and international affairs. Other programs on the North American Service are what might be termed "slices of Soviet life," which are inter-

FIGURE 6-1

РАДИО МОСКВА

RADIO MOSCOW

Dear Mr, Helms,

This verifies your report on the reception of Radio Moscow's broadcast for North America

Date November 17, 1987

Time 06:38-07:00 UTC

Frequency 7.34 MHz via Petropavlovsk-Kamchatsky.

Best wishes from Radio Moscow

MOSCOW. The Lenin Library

Государственная ордена Ленина библиотека СССР имени В. И. Ленина

Фото Е. Рябова

A QSL from Radio Moscow. Note the "rounded" frequency and the transmitter site indicated.

views (with translations) of persons from various occupations and in different parts of the Soviet Union.

Perhaps the biggest recent improvement in Radio Moscow has involved its newscasts and commentaries. No one has yet mistaken a Radio Moscow newscast for the BBC, but the level of candor and honesty has truly risen of late. You will hear frank admissions of some Soviet problems and that there are sometimes two different but equally valid ways to view a situation. While Radio Moscow is still clearly the official voice of the USSR (and, as this was being written, still closely following the existing Communist Party line), the effects of *perestroika* and *glasnost* are obvious. The Radio Moscow of 1990 is indeed greatly changed from the turgid drone of 1985, and is far more interesting and listenable as a result.

One thing that has not changed, unfortunately, is the miserable technical quality of Radio Moscow's transmissions. While signals are almost always strong, the audio quality is often poor (sounding "muffled") and it's not uncommon to hear a hum in the audio of their transmissions. The reason for this is that much of Radio Moscow's equipment is antiquated and has not been kept in proper repair; the ailments and shortages of the Soviet economy as a whole also afflict their broadcasting efforts.

Radio Moscow actively solicits reception reports, and sends out full data QSL cards in reply. QSL card designs are changed several times each year. One interesting oddity about Radio Moscow QSLs is that frequencies are specified only to two significant digits. This means a frequency such as 9615 kHz will be indicated on the QSL as "9.61 MHz." Radio Moscow also announces frequencies in the same manner, so don't assume something's wrong with your receiver's frequency readout if you find it indicating a frequency 5 kHz higher than the one announced by Radio Moscow.

Radio Moscow has been known to indicate the transmitter site used for a specific frequency on a QSL card upon request. Some SWLs like to have this done, because SWL hobby standards consider the various republic of the USSR to be separate "countries" and each transmitter site to be a separate station. However, Radio Moscow uses outdated data filed with the International Telecommunications Union to indicate transmitter locations, and there is often some doubt as to whether the site indicated is the actual transmitter location. Radio Moscow's willingness to indicate the transmitter site varies over time, although it is often closely correlated to the state of American-Soviet relations. Radio Moscow also has been known to send a pennant to regular (that is, several reports over many months) reporters. A single report to Radio Moscow is usually enough to get you on their mailing list for program schedules for the next few years.

Radio Moscow's address is simple enough: Radio Moscow, Moscow, USSR. It helps to indicate which service, such as the

North American or World Service, you're reporting on the envelope. Radio Moscow's studio address in Moscow is Pyatnitskaya Ulitsa 25; they have been known to receive American visitors with an advance appointment.

There are other international broadcasters in the USSR, with most sharing facilities with Radio Moscow. (In fact, some of these "different" stations can be heard signing on immediately after the end of a Radio Moscow transmission with no break in the carrier or change in signal strength!) At one time, these stations were little more than clones of Radio Moscow. But the tensions that began in 1990 between the constituent republics of the USSR were reflected to an extent in these stations, and it's not out of the question that some genuine competition for Radio Moscow might emerge from them.

It's unlikely that competition will come from Radio Station Peace and Progress, which bills itself as "the voice of Soviet public opinion." Supposedly operated by various Soviet cultural and social groups, it appears to be nothing more than a separate department of Radio Moscow. Radio Station Peace and Progress transmits half-hour long programs in ten languages; some languages, such as Creole and Guarani, are a bit obscure. Programs tend to be mainly news and commentaries, and have typically taken a harder anti-Western line than Radio Moscow. Interestingly, the one part of the world Radio Station Peace and Progress doesn't broadcast to is North America. However, it can usually be heard well in North America and will verify reception reports sent to Radio Station Peace and Progress, Moscow, USSR.

Some competition for Radio Moscow is possible from Radio Kiev. Kiev is located in the Ukrainian SSR, a republic that has one-fifth of the USSR's population and a disproportionate share of the Soviet Union's agricultural and industrial output. Ukrainians have long considered themselves a separate and distinct people, and 1990 saw the Ukraine declare its sovereignty and that its laws took precedence over those of the Soviet Union. While short of a full declaration of independence, these steps certainly started the Ukraine down the road to being a separate nation again. As such, Radio Kiev bears close watching. Its broadcasts in English for North America last a half-hour and are repeated twice nightly, usually at 0030 and 0300 (sometimes an hour earlier during the summer). As with Radio Moscow, Radio Kiev uses frequencies throughout the international broadcasting bands and signal strength is often excellent. The address for reception reports and the latest schedule is Radio Kiev, Radio Centre, Kiev, Ukrainian SSR, USSR.

FIGURE 6-2

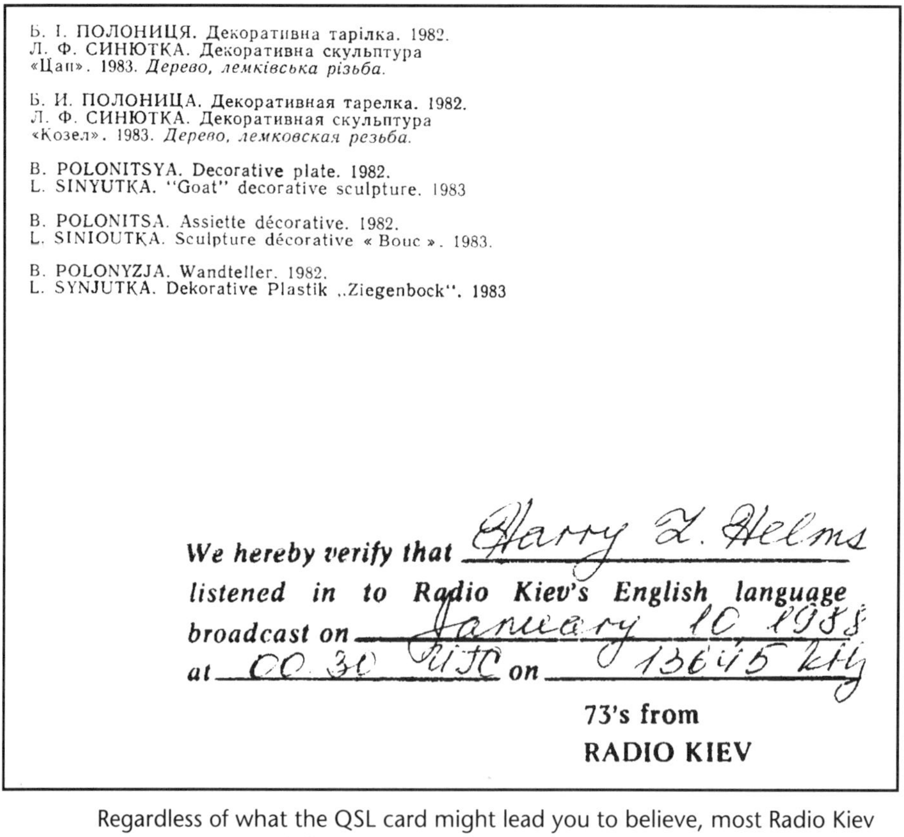

Б. І. ПОЛОНИЦЯ. Декоративна тарілка. 1982.
Л. Ф. СИНЮТКА. Декоративна скульптура «Цап». 1983. *Дерево, лемківська різьба.*

Б. И. ПОЛОНИЦА. Декоративная тарелка. 1982.
Л. Ф. СИНЮТКА. Декоративная скульптура «Козел». 1983. *Дерево, лемковская резьба.*

B. POLONITSYA. Decorative plate. 1982.
L. SINYUTKA. "Goat" decorative sculpture. 1983

B. POLONITSA. Assiette décorative. 1982.
L. SINIOUTKA. Sculpture décorative « Bouc ». 1983.

B. POLONYZJA. Wandteller. 1982.
L. SYNJUTKA. Dekorative Plastik ,,Ziegenbock". 1983

We hereby verify that Harry L. Helms ***listened in to Radio Kiev's English language broadcast on*** January 10 1988 ***at*** 00 30 UTC ***on*** 13645 kHz

73's from
RADIO KIEV

Regardless of what the QSL card might lead you to believe, most Radio Kiev transmissions aren't from the Ukraine. Instead, transmitter sites scattered across the USSR are used.

In addition to the Ukrainian SSR, other Soviet republics with their own external shortwave services in English include Uzbekistan (Radio Tashkent), Armenia (Radio Yerevan), and Lithuania (Radio Vilnius). Radio Vilnius raised the specter of what might lie in the USSR's future in 1990 with its unilateral declaration of independence. With its declaration of independence, Radio Vilnius switched from parroting Radio Moscow into a fully separate broadcaster, including strong anti-Soviet remarks in its commentaries and programs during the days following Lithuania's independence declaration. (Announcers at Radio Vilnius even began to refer to their country as "an occupied territory" and broadcast appeals for international recognition of Lithuania's independence!) Unfortunately for Radio Vilnius, its international service is relayed primarily via transmitters belonging to Radio Moscow. Abruptly, all channels used for Radio Vilnius's international programs were taken over by Radio Moscow programs. (This was later explained—not very convincingly—by Moscow officials as a "switching error.") This lasted for a couple of days, and was then replaced by another tricky move. Instead of the powerful transmitters formerly available to Radio Vilnius, weak, barely audible transmitters were used instead. While Radio Vilnius programs were still transmitted, reception was so poor that most of the programs couldn't be understood. Later, on those same frequencies, powerful transmitters could be heard with the programs of Radio Moscow and Radio Kiev. In effect, the USSR was attempting to jam a transmission *originating* from within what it claims as its territory!

The latest challenge to Radio Moscow's dominance of international broadcasting from the USSR came from a very unexpected source—Radio Station Pacific Ocean, a station located at Vladivostok that normally transmits in Russian for the Soviet fishing fleet and others in the Soviet Far East. In mid-1990, it started a brief (five to fifteen minutes long) English newscast broadcast Saturdays around 1845 UTC on 15180 kHz. This newscast was surprising because it had a strong anti-Mikhail Gorbachev slant and instead supported Boris Yeltsin, Gorbachev's main rival at the time.

At the time this is being written, the eventual outcome of events in the USSR cannot be foreseen with any certainty. What is clear, however, is that shortwave radio will allow you to literally hear history in the making,

Radio Beijing

On June 4, 1989, at 0300 UTC, listeners to the English language service of Radio Beijing heard the following announcement:

> *This is Radio Beijing. Please remember June the third 1989. The most tragic event happened in the Chinese capital, Beijing. Thousands of people, most of them innocent civilians, were killed by fully armed soldiers when they forced their way into the city. Among the killed are our colleagues at Radio Beijing. The soldiers were riding on armored vehicles and used machine guns against thousands of local residents and students who tried to block their way. When the Army convoys made the breakthrough, soldiers continued to spray their bullets indiscriminately at crowds in the street. Eyewitnesses say some armored vehicles even crushed foot soldiers who hesitated in front of resisting civilians. Radio Beijing English department deeply mourns those who died in the tragic incident, and appeals to all its listeners to join our protest for the gross violation of human rights and the most barbarous suppression of the people. Because of the abnormal situation here in Beijing, there is no other news we could bring you. We sincerely ask for your understanding, and thank you for joining us at this most tragic moment.*

The announcer has not been heard since on Radio Beijing.

This dramatic example shows how shortwave radio can put you "at ringside" when history is made. Unfortunately, it also shows how Radio Beijing, like China itself, has regressed in recent years.

In the late 1960s, Radio Beijing (then known as Peking) was a shrill, incessant source of revolution and the thoughts of Mao Tse-tung. As China began to open itself to the West, Radio Beijing began to moderate its tone. The thoughts of Chairman Mao were left back in the famous "Little Red Book" and instead programs began to deal with frank admissions of the difficulties China was undergoing in its modernization drives. Corruption and black marketeering were discussed by Radio Beijing, as were the problems of adapting Western technologies and management practices to an old and traditional culture. Since the Tien´anmen Square Massacre, however, the station has reverted to its old party-line mouthpiece ways.

Radio Beijing opens its broadcasts with an interval signal consisting of the first 19 notes from "The East is Red." This is interspersed with identification announcements in the language of the target area. The first item on broadcasts is a newscast, followed by a commentary on the news. A significant number of items deal with Chinese politics and events, and these can be a rich source of material on Chinese affairs and politics.

Radio Beijing still taps into China's long, rich cultural history for programming material. There are programs based on Chinese folk tales, archeology, and history (such as the origin of paper in China). There is still much contemporary and traditional Chinese music; one doesn't have to understand Chinese to enjoy and appreciate the voices.

Unfortunately, the non-musical, non-traditional programs of Radio Beijing have backslid since June of 1989. There is far more propaganda and stridency, with a strong air of unreality. A constant effort is made to deny that anything significant happened in June, 1989; the strain of having to broadcast such lies must be a major burden on the Radio Beijing staff.

For many years, Radio Beijing was difficult to receive in North America since most signal paths had to cross the North Polar regions. Now Radio Beijing is relayed via Swiss Radio International, Radio Canada International, and Radio France International's station in French Guiana. Radio Beijing also has a relay station in the African nation of Mali. This relay network means that Radio Beijing can now be heard reliably in North America.

Radio Beijing welcomes letters and reception reports from SWLs. In addition to colorful QSL cards, you can expect to receive magazines, booklets, calendars, and other items if you're a regular reporter to them. Their address is simply Radio

FIGURE 6-3

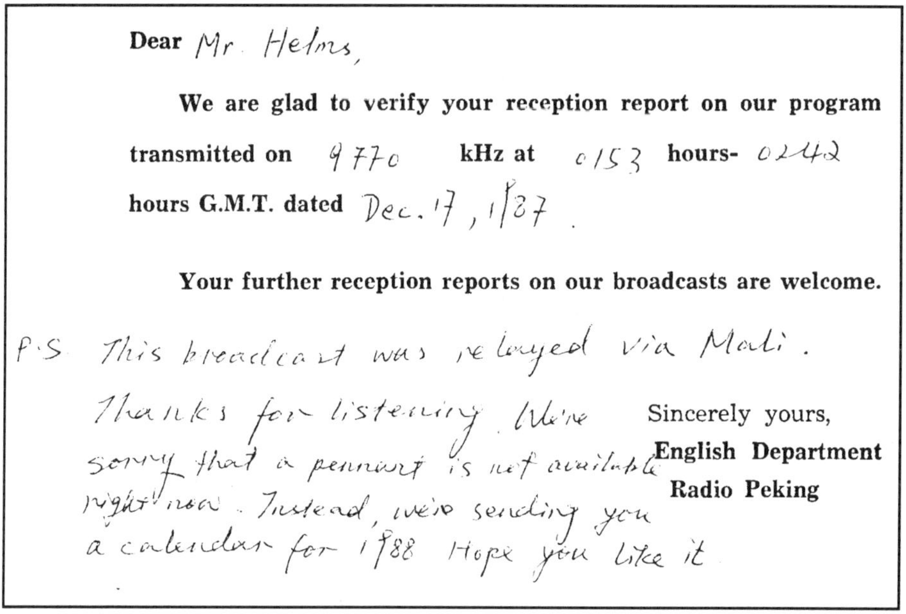

Dear Mr. Helms,

We are glad to verify your reception report on our program transmitted on 9770 kHz at 0153 hours- 0242 hours G.M.T. dated Dec. 17, 1987.

Your further reception reports on our broadcasts are welcome.

P.S. This broadcast was relayed via Mali. Thanks for listening. We're sorry that a pennant is not available right now. Instead, we're sending you a calendar for 1988 Hope you like it

Sincerely yours,
English Department
Radio Peking

The handwritten note confirms reception of Radio Beijing via Mali. At the time of this reception, the relay had just come on the air and there was still confusion over exactly where it was!

Beijing, Fu Xin Men, Beijing, People's Republic of China. Use the full name of the country—Chinese postal authorities have been known to return letters simply addressed to "China."

The future of China was uncertain as this was being written, and the possibility of further upheavals is great. Radio Beijing might have more dramatic moments for listeners in the future.

Private American International Shortwave Stations

Domestic shortwave broadcasting has never existed to any real extent in the United States, mainly because there are few areas (with the exception of northern Alaska) where several AM and FM stations cannot be received easily. Moreover, the difficulty of tuning and using older shortwave radios helped ensure that people would be content with what they could find on their local broadcasters. The FCC eventually adopted a policy of licensing private shortwave broadcasters only when they targeted international audiences.

For many years, most private shortwave stations from the United States have been religious broadcasters. Several of these are still active today, and call signs such as WYFR, WINB, WHRI, and WMLK will soon become familiar to you. Unfortunately, most of these stations broadcast the same "canned" religious programs (along with the constant appeals for money) that you can already hear on AM and FM stations.

By the late 1970s, some in the United States began to seriously consider the possibilities of commercial international broadcasters. The growing affluence of the rest of the world meant advertising possibilities were greater, and the revolution in SW radio technology meant finding a specific frequency or station was no longer such a chore. The first station to try commercial international broadcasting was WRNO in New Orleans. WRNO was already a successful FM station with a rock and roll format, and its plan for international broadcasting was simple: sell some air time to religious broadcasters to provide an income "floor" and then "simulcast" the FM station the rest of the day. WRNO began by supplementing its offerings with a DX program hosted by well-known SWL author and publisher Glenn Hauser as well as worldwide call-in request programs. Apparently, the WRNO experiment has not been as successful as first hoped; the station's programming today consists almost exclusively of FM relays and religion with hardly any reference to shortwave. This has not deterred others from trying their luck, however. One recent entry is KUSW from Salt Lake City, which broadcasts (at the time this book was being prepared) an extensive pop music format and several features aimed at SWLs.

One effort at commercial international broadcasting in the United States deserves special mention if for no other reason than its audacity and sheer *chutzpah*. In the 1986 edition of the *World Radio TV Handbook*, readers were startled to see a survey card bound into the book for a proposed station called "NDXE Worldwide" to be located in Opelika, Alabama. Their plans called for 24-hour broadcasting in English, Spanish, French, Portuguese, and Japanese, with news, sports, music, and (to quote their literature) "in-depth coverage of financial, scientific, and technological developments." The station's backers circulated plans for a 500 KW transmitting facility and even talked of AM stereo broadcasting on shortwave. Even as late as 1989, the NDXE backers were still trying to drum up interest in their proposal. However, they never put their grandiose scheme in action, and apparently never got remotely close to arranging the necessary financial backing for their venture. Tired of the entire affair, the FCC summarily rejected the NDXE group's application for a license in 1989 and put an end to the matter.

A much more worthy entrant to the private international broadcasting scene was the introduction of a "world service" by the *Christian Science Monitor*. This service consists of news, news commentaries and analysis, and features related to current

FIGURE 6-4

QSL from WCSN, Scotts Corner, Maine.

events. Like the newspaper, this service is a bit dry and dull but also scrupulously fair and accurate. There is also none of the heavy-handed appeals for money that spoil other American religious shortwave broadcasters. Currently, the World Service of the *Christian Science Monitor* is transmitted in the United States over WCSN in Maine and WSHB in South Carolina, and via KYOI on the island of Saipan in the U. S. territory of the Mariana Islands. All of these stations can be heard well in North America.

Radio France Internationale

Radio France Internationale (RFI) recently made a major innovation in its programming to North America—it added an hour of English each day, supplementing the several hours of

French broadcasts. For many years, France resolutely refused to broadcast in English to North America. France did broadcast extensively in English to Africa, and even had fifteen minutes of English daily in its broadcasts to South America. The reason behind this *de facto* boycott of North American English listeners was never explained; probably the explanation lies somewhere deep within the French psyche.

Radio France Internationale has long had a particular interest in programming for Africa, a situation not too surprising in view of France's long colonial history on that continent and continuing relationships with many of its former colonies. RFI's programming to Africa deals more with the target area than France itself. For years, it broadcast an hour-long English program known as "Paris Calling Africa" which dealt exclusively with African news and events. This was later changed to a format of alternating half-hours of French and English programs transmitted to Africa. The emphasis is still on African news, however, and RFI's programs to Africa are a key source for information on African affairs.

RFI's English programs to Africa and other areas have a delightful spontaneity to them. Many are apparently done live or in "one take." Thus, you'll sometimes hear announcers abruptly interrupt their newscasts for a "hot item" which just arrived. (The importance of these items escapes me; usually, they seem trivial.) At other times, you can hear announcers fumble with papers or frantically search for missing items; I once heard an RFI announcer ask "I can't find it! Where is it?" in the middle of a newscast! Compared to the methodical, carefully rehearsed presentations of most other international broadcasters, Radio France Internationale seems very human and real. In addition to gaffes, RFI also features reviews of the French press (invariably the Paris papers) in its English language programs along with plenty of French music.

French music can be enjoyed, even if you don't speak French, during RFI's French broadcasts. Moreover, if you're studying

French in school (or want to resurrect what you studied years ago), then RFI is an excellent source of contemporary idiomatic French. (It's surprising that more schools and educators haven't taken advantage of the enormous potential of shortwave radio in language study.)

Radio France Internationale recently made a major effort to improve reception in North and South America when it opened a relay station in French Guiana, a French colony located on the northern coast of South America. This station puts powerful signals in North America, and also serves as a relay site for other stations such as Radio Beijing.

While Radio France Internationale does not go out of its way to solicit letters and reception reports, it does send out colorful

FIGURE 6-5

Radio France Internationale's relay station at French Guiana is shown on this card verifying reception via it.

QSL cards in response to correct reports. These can be addressed to Radio France Internationale, P. O. Box 9516, 75762 Paris Cedex 16, France. As an advertisement for RFI in a shortwave publication once put it, "Wherever you are—provided you know how to use your receiver—you can hear Radio France Internationale." That line sums up RFI and its attitude in a nutshell!

Evangelical Broadcasting: HCJB and Trans World Radio

As mentioned earlier, various Christian evangelical groups make extensive use of shortwave broadcasting. This is in marked contrast to other major religious groups; for example, despite the massive wealth accumulated in recent years by various Islamic countries, there has been no effort to establish stations and broadcasting organizations to further the spread of Islam. The effectiveness of evangelical broadcasting is hard to measure and widely debated. Regardless of its actual impact, evangelical broadcasting has steadily grown over the past two decades, and more religious organizations seem ready to devote time and resources to broadcasting. Two "prototypes" of international broadcasting are station HCJB, "the Voice of the Andes" in Quito, Ecuador, and Trans World Radio (TWR), a New Jersey organization which operates stations from such locations as Monaco and Guam.

HCJB is operated by World Radio Missionary Fellowship, a group headquartered in Opa Locka, Florida. While the call letters are drawn from the block reserved for Ecuador, the selection was not accidental; they stand for *H*eralding *C*hrist *J*esus's *B*lessings. The scope of HCJB's operations is greater than some government-financed broadcasters, and include operations in such languages as Russian and Japanese. Transmitting facilities include a 500 KW unit, and the station has even constructed its own hydroelectric station to provide power for its facilities. HCJB also operates a transmitter on 690 kHz which broadcasts in Spanish and local Indian languages.

HCJB has also made a strenuous effort to maintain good relations with SWLs, including a weekly program specifically for SWLs titled "DX Party Line." HCJB welcomes listener mail and reception reports, and often issues series of colorful QSL cards for collectors.

If you want to learn about the politics and culture of Ecuador and South America, you'll be disappointed by HCJB. Politics is strictly eschewed, and content relating to Ecuador and South America is subordinate to the prime mission of evangelism. You'll hear many programs, such as "Guide to Family Living," which are similar to those found on your local AM and FM religious broadcasters. Even those programs which are ostensibly not religious, such as "DX Party Line," manage to work in some moments connected with evangelical Christianity. The approach is generally "soft sell," but the prime purpose behind HCJB is to convert listeners to the religious viewpoints of the station's sponsors. There is nothing sinister in this, of course, but it does explain why its program content is sometimes indifferent to Ecuador and South America. On the other hand, HCJB's 690 kHz station does have a strong "service orientation" for its local audience, with a greater proportion of educational and cultural content.

HCJB welcomes reports to Casilla 691, Quito, Ecuador; return postage is appreciated if a QSL card is desired. In addition to a QSL card, SWLs can expect to receive a copy of HCJB's latest program schedule and a religious tract or two. While transmission schedules and frequencies do change seasonally, the powerful signals from HCJB are easy to spot in North America.

Trans World Radio is exclusively an evangelical broadcaster, using several different transmitter sites. One is located in Monaco, a tiny nation between France and Italy, and transmits religious programs in a variety of European languages.

Shortwave transmitter powers up to 500 KW are used, but all programs are intended only for Europe. However, it usually schedules English programs for Europe between approximately 0500 to 1100 UTC in the 41-meter, 31-meter, and 25-meter bands, and these can be received in North America. You won't hear any news or information about Monaco (or any other country TWR broadcasts from) on TWR; all programs are strictly religious. In fact, unless you're interested in the religious programming itself, there will be little reason for you to become a regular listener to any TWR station.

The prime reason most SWLs tune to TWR at Monaco is to secure a QSL card for Monaco. In English, you'll note the station identifies itself as transmitting from Monte Carlo instead of Monaco; Monte Carlo is Monaco's capital (for all practical purposes, Monte Carlo *is* Monaco!) and home to the casinos that draw tourists and produce wealth for the tiny principality. You can send your reception reports to Trans World Radio, B. P. 349, Monte Carlo-98007, Monaco.

Incidentally, the question of whether you can really hear Monaco via TWR is disputed by some SWLs since the transmitters for TWR are actually on French soil. A small (but vocal) minority in the SWLing hobby maintains that all TWR/Monaco broadcasts should be counted as being from France rather than Monaco. Most SWLs, however, continue to use the location where a station is licensed or authorized as the criterion for determining transmitter location unless the geographic separation between the authorized location and transmitter site is significant. In the case of Monaco, the separation is minor. Is this a logical approach? Maybe not, but then not everything in life makes sense....

The TWR outlet you're most likely to hear is the one located on the island of Bonaire in the Netherlands Antilles, a group of islands off the coast of Venezuela. In addition to SW, TWR also transmits on the AM broadcast band using a 500 KW transmitter on 800 kHz. English is often scheduled from 0230 to 0500

UTC, and listeners in the eastern two-thirds of North America can often hear TWR easily on 800 kHz if conditions are auroral. As you might expect, TWR/Bonaire broadcasts mainly in the languages spoken in the Americas, such as Spanish, English, and Portuguese. However, they have also used Russian and German. On shortwave, English is often scheduled around 0400 to 0500 and 1100 to 1230 UTC; frequencies in the 25-meter band are commonly used. All transmissions are clearly identified as coming from Bonaire, and TWR does QSL correct reception reports sent to Trans World Radio, Bonaire, Netherlands Antilles. You may recognize the station by its interval signal, which consists of the opening notes of "Stand Up, Stand Up for Jesus" played instrumentally.

Another TWR outlet operates from the U.S. territory of Guam. Since this station operates from the jurisdiction of the United States, it has call letters—KTWR. The shortwave service is intended for reception in Asia, and makes extensive use of Asian languages such as Bengali, Cantonese, Indonesian, Japanese, and Tamil. There are also some English programs.

Unlike some other TWR outlets, KTWR has broadcast features of interest to SWLs and DXers such as "DX Listener's Log" in both English and Japanese. Reception reports and requests for program schedules should go to KTWR, P. O. Box CC, Agana, Guam, 96910. Their interval signal is the first few notes of "We've a Story to Tell to the Nations" played on an organ.

A more difficult TWR outlet to hear in North America is their station located in Swaziland. This site is used for programming to Africa, and its times and frequencies of operation are not well-suited for North American reception. Moreover, the most powerful transmitter used in Swaziland is only 100 KW and, more typically, 25 KW is used. Several frequencies is the 90-meter and 60-meter tropical broadcasting bands are employed. Besides English and French, you can hear such exotic languages as Chewa, Lingala, Ndebele, Sotho, Tshwa, and Zulu

FIGURE 6-6

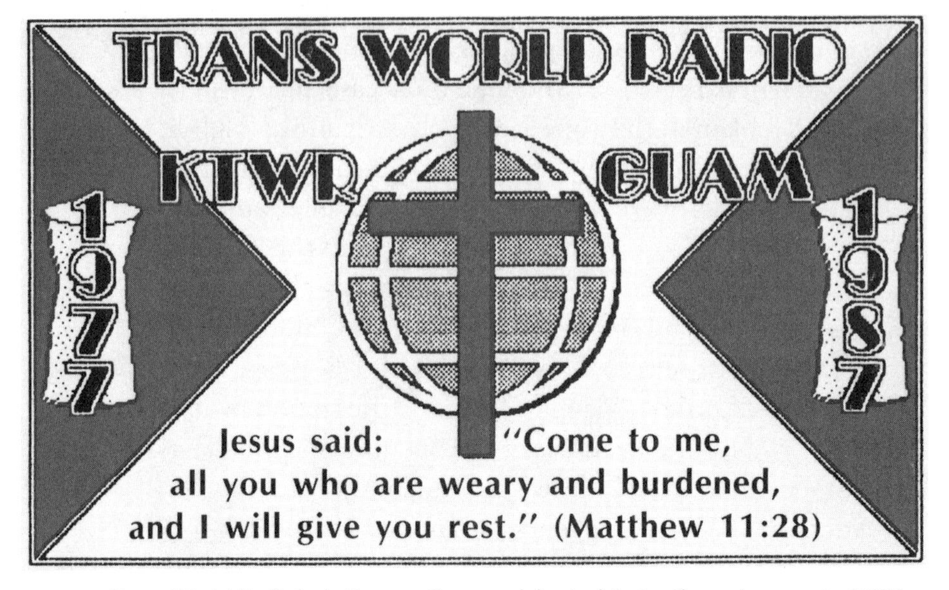

Trans World Radio's station on Guam celebrated its tenth anniversary in 1987 with a special QSL card.

over TWR/Swaziland. You may find it easier to identify the station by its interval signal, which is the same as that used for KTWR but played on bells. The address for reception reports and program schedules is P. O. Box 64, Manzini, Swaziland.

TWR also maintains another transmitter location in Sri Lanka (formerly known as Ceylon) and also broadcasts over Radio Monte Carlo's relay station on the island of Cyprus. However, both of these operations exclusively take place on the AM broadcast band, and reception in North America is unlikely.

As mentioned at the beginning of this section, religious broadcasting activity has been increasing rapidly in recent years. Regardless of one's opinions about such efforts and programs, they have been the force behind the establishment of several new international broadcasting stations in the United States. These stations show no sign of lessening their impact on SW radio in the foreseeable future.

Vatican Radio

One might think Vatican Radio would be in the same category as HCJB and TWR. However, HCJB and TWR broadcast to convert listeners while Vatican Radio, in effect, broadcasts to the already converted. Vatican Radio's programs are usually brief—a half hour or less—and deal primarily with Church affairs and news. Programs are presented in over thirty languages, including Esperanto. Other broadcasts include celebration of the Mass in various languages, including Polish and Latin.

Vatican Radio's broadcast to North America open with "Christus Vincit" played on a celeste followed by the phrase "Laudetur Jesus Christus. Praised be Jesus Christ. This is Vatican Radio." This is usually followed by news and commentary, and then a feature dealing with some aspect of the Church or Vatican City. Interviews with American visitors to Vatican City are a frequent topic. At other times, Vatican Radio broadcasts live addresses by the Pope. The normal schedule of Vatican Radio goes out the window when the Papacy is vacant; the first word that a new Pope has been elected often comes from Vatican Radio. (In fact, published reports have stated that Vatican Radio had a "spy" with a concealed transmitter in the conclaves that elected Popes John Paul I and John Paul II to ensure they would be the first to announce the new Pope.)

There is usually one daily transmission for North America beginning at 0050 UTC; one frequency each in the 49-meter, 31-meter, and 25-meter bands is typically used. Transmitter powers range up to 500 KW, and signal strength is usually excellent.

Vatican Radio welcomes listener reports and responds with QSL cards that are changed on a regular basis. They also send out a multilingual program schedule to reporters. The address is simply Vatican Radio, Vatican City.

The status of Vatican City as a separate radio "country" is

questioned by some of the same SWLs who question the status of TWR in Monaco. Vatican City itself is obviously too small to contain a major broadcasting facility; in fact, the actual transmitting site itself is located some distance from Rome. However, under an agreement between Vatican City and Italy, the transmitter site has extraterritorial status from Italy and is administered by Vatican City. A few SWLs disregard this and insist that all Vatican City transmissions originate from Italy. However, the majority of SWLs consider the agreement valid and recognize Vatican Radio's transmitter site as separate from Italy.

Radio Nederland

The Netherlands is not a large country in terms of geography, economic power, or military power, but it's a giant when it comes to "information power" thanks to Radio Nederland. Radio Nederland has long been a favorite of SWLs. Its programs are consistently entertaining and informative, and it has shown a special interest in SWLing for years. Moreover, it is a technically innovative broadcaster. Radio Nederland was one of the first stations to experiment with SSB transmission of programs and its transmitter facilities, including relay facilities on Madagascar and Bonaire, are among the most "state of the art" of any international broadcaster. The Netherlands's commitment to international broadcasting was underscored in 1985, when Radio Nederland placed in operation a massive new transmitter site in the Flevopolder area of the country. The land was reclaimed from the sea, and the facility is actually *below* sea level.

Radio Nederland only uses a handful of languages, and concentrates its efforts into delivering good signals and good programs in those languages. All of its programs begin with news and a commentary followed by various features. As befits an open, tolerant society whose prosperity has traditionally been based upon international commerce, Radio Nederland is very internationalist in its outlook. Radio Nederland features the usual programs (life in the Netherlands, interviews, pop music, listener mailbag, etc.) along with some offbeat features such as the first program on an international shortwave broadcaster specifically for gay and lesbian SWLs! Radio Nederland seems to be one of the few international broadcasters to have grasped that good, interesting programs are the way to keep people tuning in.

Radio Nederland various relay sites can make it difficult to determine just which location you're tuned to. Fortunately, transmitter sites are indicated in Radio Nederland's program schedule and are announced when each transmitter signs on or a regional service (such as to North America) begins. Each site tends to serve adjacent areas. For example, Bonaire is primarily used for broadcasts to the Americas, while Madagascar is used for transmissions to Africa and Asia. For the evening broadcasts to North America, most transmissions are from Bonaire,

FIGURE 6-7

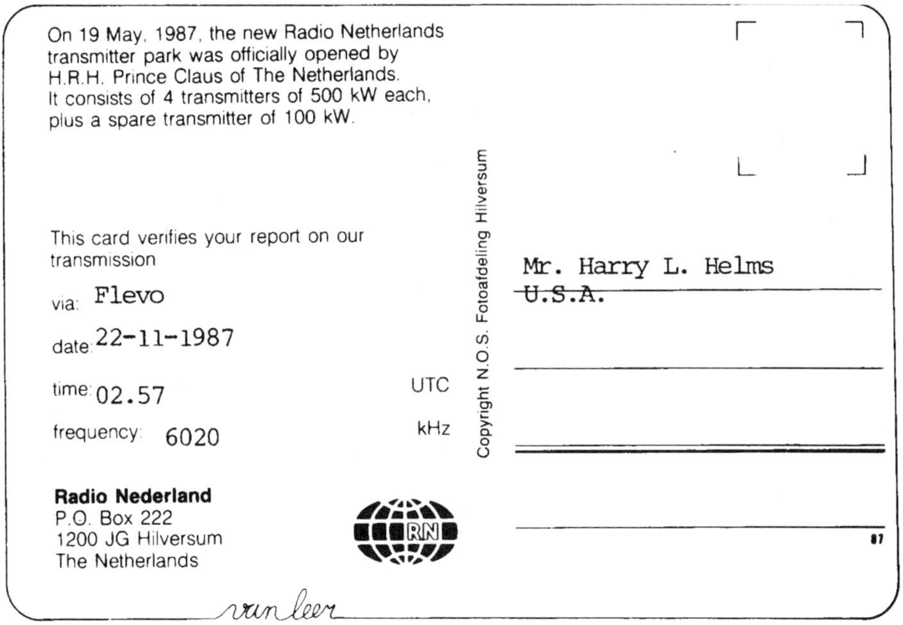

On 19 May, 1987, the new Radio Netherlands transmitter park was officially opened by H.R.H. Prince Claus of The Netherlands. It consists of 4 transmitters of 500 kW each, plus a spare transmitter of 100 kW.

This card verifies your report on our transmission

via: Flevo

date: 22-11-1987

time: 02.57 UTC

frequency: 6020 kHz

Copyright N.O.S. Fotoafdeling Hilversum

Mr. Harry L. Helms
~~U.S.A.~~

Radio Nederland
P.O. Box 222
1200 JG Hilversum
The Netherlands

RN

van leer

87

Radio Nederland is sometimes difficult to hear in North America directly instead of from Bonaire!

although Radio Nederland often uses at least one frequency from Flevopolder. You won't have much difficulty recognizing the signals from Bonaire; they are quite strong and have little fading, since most of North America is within "single hop" range of Bonaire.

As you might expect, Radio Nederland also likes letters and reports from SWLs. They prefer comments on programming more than data on reception, as they have a network of regular monitors worldwide who contribute information. However, they do send out colorful QSL cards in response to correct reception reports (although they do ask you to keep your requests to one QSL card per month to help conserve their resources). You can send your reports and request a program schedule by writing Radio Nederland at P. O. Box 222, 1200 JG Hilversum, The Netherlands.

Radio Nederland uses a beautiful interval signal consisting of an old Dutch folk tune played on a carillon and celeste. If you're like many SWLs, that sound—and Radio Nederland—will soon become quite familiar to you.

The Voice of America

It might seem that the Voice of America (VOA) wouldn't be of much interest to American SWLs. Such isn't the case. It's always interesting to see how one's tax revenues are bring spent and how the United States tries to present itself abroad.

The Voice of America is the broadcast division of the United States Information Agency (USIA); USIA also distributes print, audio, and video materials about the United States throughout the world. The USIA is headed by a director appointed by the President; this has resulted in complaints by some VOA staff members and other observers that such appointees have tried to influence the editorial processes of the VOA and other USIA functions to reflect favorably upon the administration and political party in power.

The VOA is a massive broadcasting organization, currently producing programming in over forty languages. In the United States itself, transmitter sites at Greenville, North Carolina; Bethany, Ohio; Delano, California; and Dixon, California are used. The Greenville, North Carolina facility is the main VOA transmitting site in the United States, and transmitter powers up to 500 KW are used from there. The other domestic sites have transmitters rated at 250 KW. There is another VOA station, at Marathon, Florida, which operates only on the AM broadcast band for programs to Cuba.

The Voice of America uses an extensive network of relay stations. SW transmitting facilities are maintained by the VOA in Germany, Greece, Liberia, Morocco, the Philippines, and Sri Lanka. Other relay sites are planned for Israel and Thailand. The VOA is also relayed in Britain by the BBC under a reciprocal agreement (in turn, the VOA relays the BBC in the United States) and has also been relayed by Radiobras, the Brazilian government broadcasting service. In addition, the VOA maintains AM broadcast band relay stations in Antigua, Belize, and Botswana. Identifying the various transmitter sites is easy if you tune to the start of a VOA transmission; the sign-on announcement includes the transmitter location. The VOA uses several different instrumental versions of "Yankee Doodle" for an interval signal.

Many SWLs (myself included) find the English language programs of the VOA dull compared to other international broadcasters. Part of this arises from the desire of VOA staff to be as impartial and objective as possible in both the selection and treatment of topics broadcast. However, the results too often sound like "programming by committee." There's little that might be offensive, but unfortunately just as little that might be really interesting.

The VOA's newscasts are not as reliable or as well-respected as those of the BBC, but are generally honest and reasonably accurate. It also has news and events programs covering the target areas it broadcasts to, such as "Caribbean Report" and "Nightline Africa." The VOA also broadcasts considerable American music—including America's two original musical forms, jazz and rock and roll—on programs such as "Music Now" and "Music USA."

The Voice of America presents some newscasts and programs in what is known as "special English." Such programs use a limited vocabulary of common English words and verb tenses; the announcers speak very slowly and enunciate carefully to allow persons with a limited command of English to both understand VOA newscasts and improve their English skills.

The VOA does not neglect SWLs. The VOA is an excellent verifier of reception reports. Voice of America receptions are not always routine. For example, the Sri Lanka relay uses transmitters of only 35 KW, and reception in North America is

FIGURE 6-8

VOA

THE VOICE OF AMERICA
Washington, D.C., U.S.A. 20547

VERIFICATION CARD

We are pleased to confirm your reception of our broadcast on

February 2, 1988
3990 kHz
0600-0616
Monrovia, Liberia

Space Shuttle Challenger blasting-off from the launch pad at the Kennedy Space Center on its maiden flight on April 4, 1983.

Thank you for your interest in VOA. We hope you will continue to enjoy our programs.

The Voice of America can sometimes be a challenging DX target, as this reception on 3990 kHz from Liberia illustrates.

a real challenge even for experienced SWLs. The VOA's British relay is the easiest way to get a "full data" QSL for that country, and the Morocco, Liberia, and Philippines relays are the easiest way to hear and verify those countries. Reception reports and requests for program schedules go to Voice of America, U.S. Information Agency, Washington, DC, 20547.

For most American SWLs, SWLing lets us know how other countries view us. The Voice of America lets us see how we view ourselves—at least "officially." And, unlike other SW broadcasters, if you don't like the VOA, you can always complain to your Congressional representatives!

The Changing Face of International Broadcasting

While some international broadcasters are busy expanding their facilities and scope of operations, others are scaling back. Radio New Zealand International and Radio RSA, "the Voice of South Africa," are two cases in point.

For years, Radio New Zealand was a favorite of SWLs worldwide, largely because of the SWL programs hosted by well-known DXer Arthur Cushen. By the 1970s, however, Radio New Zealand was awash in problems. Its transmitters were only 7500 watts, and were badly squeezed by the 250 and 500 KW transmitters being put into use by other international broadcasters. Listeners began tuning out Radio New Zealand in favor of stronger, more easily heard broadcasters. Serious discussion took place within the nation over whether Radio New Zealand should be discontinued altogether. Others in the New Zealand government favored closing down the existing facilities and simply renting air time from the facilities of Radio Australia.

Fortunately for SWLs, the decision was made to expand New Zealand's shortwave broadcasting capabilities. New 100 KW transmitting facilities were installed on New Zealand's North Island, and in early 1990 the service was renamed Radio New

FIGURE 6-9

Radio New Zealand International celebrated its new facilities by sending this station pennant to listeners.

Zealand International. The new service was an immediate hit with listeners, keeping a refreshing, low-key style and providing superb coverage of events in the southern Pacific. Like the Netherlands, New Zealand now seems determined to play a major role as an "information power."

South Africa went in the opposite direction in 1990. Radio RSA began in early 1967 as an effort to cultivate good will for South Africa and offset the negative international image it had due to its apartheid policies. Some of these efforts were creative. One was an international telephone call-in program held on New Year's Eve; Radio RSA then tried very subtly to interpret the responses as being some sort of expression of support for Radio RSA and the South African government. Letters to the station from listeners were also interpreted in a similar fashion. Radio RSA was also famed for sending out lavish, full-color QSL cards and station pennants to listeners.

But a budget crisis, brought on in part as a result of international economic sanctions against South Africa, took its toll on

FIGURE 6-10

Radio RSA was noted for its stunning QSL cards depicting African scenes.

Radio RSA. Deciding that an international broadcasting effort of such scope was too expensive for the benefits it brought, Radio RSA dropped all of its international services except those to the rest of Africa as of May 1, 1990. The last international service broadcast concluded with *Auld Lang Syne*. Reception is still possible in North America on remaining frequencies, but with much more difficulty. While it's difficult to mourn the loss of a propaganda effort whose mission was to defend indefensible policies, the loss of a major shortwave voice from South Africa hurts at a time when South Africa is poised on the brink of major change. In the meantime, Radio RSA is rumored to be offering its transmitter facilities as a relay site for other stations.

The collapse of communist regimes in eastern Europe during 1989 had a major impact on broadcasting from those nations. Perhaps the most abrupt reversal was done by Romania's Radio

Bucharest. As the last holdout among the hardline regimes, Radio Bucharest steadfastly opposed all changes to the status quo. After the ouster of Ceausescu, the station *apologized* to listeners for giving misleading news and information in the past! In Czechoslovakia, Radio Prague shut down their external broadcasting services for a couple of months to purge all communists who worked there.

The political changes in Europe closed a major shortwave broadcaster. For years, Radio Berlin International was the voice of the German Democratic Republic. Its main "competition" was West Germany's Deutsche Welle, which emerged, as West Germany prospered, as the major rival to the BBC as Europe's dominant shortwave broadcaster. As German reunification progressed in 1990, the future of Radio Berlin International

FIGURE 6-11

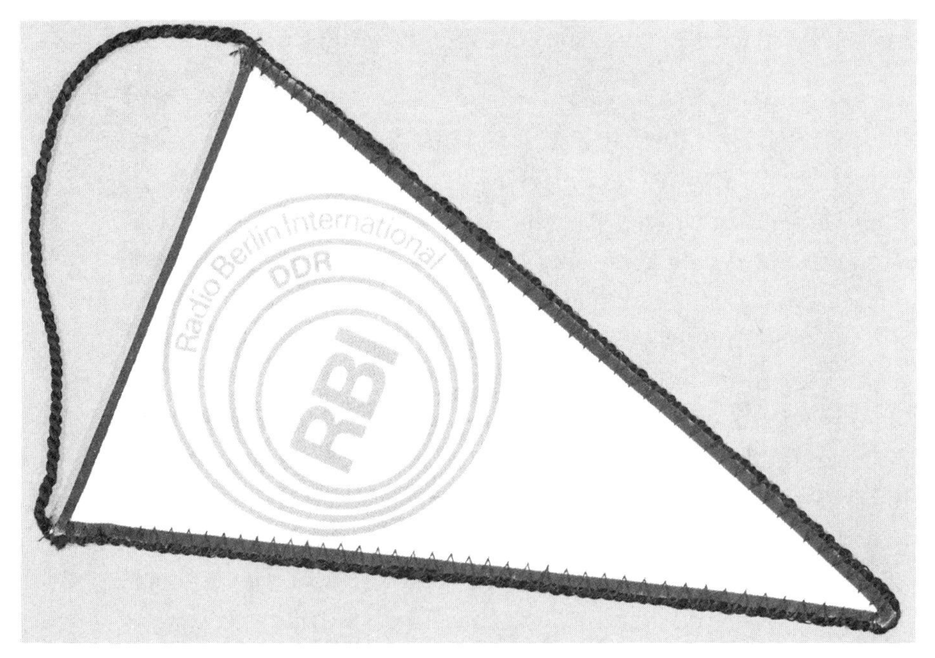

Radio Berlin International is just a memory now—and pennants such as this one are collectors items!

became clouded. Finally, on October 3, 1990, Radio Berlin International left the air forever.

If you're new to shortwave radio, you missed a lot of exciting listening from international broadcasters in 1989 and 1990. But the world is dynamic, and there are bound to be more memorable listening experiences in the years ahead!

Beyond News and Commentaries

Certainly some of the big attractions of international broadcasting are the newscasts and commentaries of the various stations. If you have an interest in international politics and events, these let you get a better insight into various countries than domestic media alone provide.

But there's more to international shortwave broadcasting than just news. For example, several stations offer "by radio" courses in the language(s) of their countries; in the past, Radio Japan and Radio Nederland have conducted such courses. Moreover, there's an endless supply of practice material in various foreign languages on international shortwave broadcasters. For example, the Voice of America and BBC let you practice everything from Arabic to Swahili.

Folk and indigenous music of various countries can be collected via tape recordings. This can be a bit frustrating, since it often takes exceptional conditions to produce signal levels worth recording. But if you're a musicologist, you'll find this a quick way to build an outstanding library of world music.

One reason some SWLs report reception to international broadcasters is to collect the stamps often used on the reply envelopes. Some broadcasters, such as Radio Moscow, make it a habit of using the latest issues on their replies to listeners. It's no surprise that stamp collecting is the second hobby of several SWLs!

This chapter has been little more than a sample of the variety of international broadcasting stations you can hear. The latest frequencies and operating schedules can be found in such publications as the annual *Passport to World Band Radio*, SWL club bulletins, and magazines like *Popular Communications* and *Monitoring Times*. To get on the mailing list for program schedules, you can send a request to the addresses in table 6-1.

TABLE 6-1
Addresses of Other Major International Broadcasters

Argentina	RAE English Services, Casilla 555, 1000 Buenos Aires
Australia	Radio Australia, GPO Box 428G, Melbourne 3001
Austria	Radio Austria International, A-1136 Vienna
Belgium	RTBF International Services, Box 202, B-1040 Brussels
Canada	RCI, Box 6000, Montreal, Quebec H3C 3A8
Cuba	Radio Havana, Apartado 7026, Havana
Egypt	Egyptian Radio & TV, P. O. Box 1186, Cairo
Finland	Radio Finland, Box 10, SF-00241 Helsinki 24
Israel	Kol Israel, Overseas Services, Box 1082, Jerusalem
Italy	RAI English Service, Viale Mazzini 14, 00195 Rome
Japan	Radio Japan, NHK, 2-2-1 Jinnan, Shibuya-ku, Tokyo
Norway	Radio Norway International, N-Oslo 3
Portugal	Radio Portugal, Av. Eng. Duarte Pacheco 5, 1000 Lisboa
Spain	Spanish Foreign radio, Partado 156.202, 28080 Madrid
Sweden	Radio Sweden International, S-10510 Stockholm
Switzerland	Swiss Radio International, Giacomettistrasse 1, CH-3000 Berne
Turkey	Voice of Turkey, P. O. Box 333, Yenisehir, Ankara

CHAPTER SEVEN

Domestic Shortwave Broadcasting

DOMESTIC SHORTWAVE BROADCASTING offers a stark contrast to the powerful signals and carefully polished programs of international broadcasters. Domestic SW stations are intended for reception within the country where the stations are located, and generally couldn't care less about any audience outside their country. Domestic SW stations are more challenging, and often difficult, listening fare. But many SWLs find them more rewarding in terms of programming, and DXers find the ultimate tests of their skill among them.

You'll probably get a little discouraged when you first try to hear the stations mentioned in this chapter. For starters, you won't hear the loud, thumping signals like those from international broadcasters. Instead of the 500 KW transmitter powers used by international stations, domestic shortwave stations might use only a few hundreds of watts. Most domestic stations operate on lower frequencies in the 60-meter or lower broadcasting bands, meaning they operate well below the MUF for a given path and signals levels are lower. Since domestic SW stations operate for the benefit of a local audience, they are on the air when their local audience is awake, not when propagation is best to a point thousands of miles away. All these factors mean that even die-hard SWLs are forced to become DXers when tuning domestic SW stations. And even if you can hear the signal, understanding it can be a big problem; the studio equipment (microphones, etc.) used by these stations is often not the best and distorted, noisy audio is not uncommon.

There's also the language barrier. You'll hear little English on domestic SW stations. Instead, expect to hear a lot of Spanish, French, and Portuguese. You'll also hear Indonesian, Tahitian, Lingala, Swahili, Hindustani, Malay, and other languages used by the populations the stations serve. Trying to identify a given station can be difficult, especially in cases where there are several stations broadcasting in the same language on the same frequency.

If listening to domestic SW broadcasters is so much trouble, why bother???

I suppose it's because this is where some of the real romance of shortwave comes into play. When you listen to domestic SW stations, you make a direct connection to another people and their culture. There's something beguiling about listening to exotic languages and music in the static. For a few moments, you are sharing something with people in a remote African village or along the Siberian frontier. You'll hear Chinese operas and the so-called "high life" music popular in Africa. You'll be amazed (or dismayed) at how American culture has insinuated itself into that of other nations; a tribal chant may be followed by a Michael Jackson tune. You'll discover that "Jungle Oats" is a popular breakfast cereal in South Africa while "Inca Cola" is a favorite of Peruvians. And you don't have to understand the language to get a kick out of hearing a soccer match from a Latin American nation as the announcers sound as if they're on the verge of a heart attack. Unlike international broadcasting, which tries to adapt itself to the culture of its target area, domestic SW stations force you to adapt yourself to their culture. And in the process several good things happen.

One of them is language knowledge. You won't get fluent in a bunch of languages, but you'll soon be able to recognize different languages when you hear them. You *will* learn how to say "this is radio station. . . ." in several different languages, and you'll also pick up several additional words and phrases. If you're studying a foreign language, or want to keep your knowledge of one fresh, it's hard to beat the practice material provided by domestic SW stations.

You'll become more musically aware. Ever heard authentic South Seas music played on slide guitars? Or how about Mexico's "ranchero" music, with its strong similarities to traditional American country and western? And it won't be long before you'll know when you hear "merengue" music that you're probably listening to a station in the Dominican Republic.

If you get interested in collecting QSLs from the stations you hear, you'll be especially eager to write domestic SW stations. It's always exciting to receive a letter with exotic stamps and an unusual postmark. Instead of the mass-processed replies from international broadcasters, you'll sometimes get long, genuinely personal replies from such stations. You get the feeling that your letter telling them they were heard far away was the highlight of the day at the station. Even printed QSL cards often come with handwritten notes on them. Cancelled postage stamps, stickers, and even pennants might be received from domestic broadcasters.

Domestic stations change frequencies and hours of operation much less often than international SW broadcasters. In fact, many stations have operated on the same frequencies for decades. Seasonal frequency changes common among international broadcasters are almost unknown among domestic broadcasters. When a change is made in a station's frequency, it tends to be relatively permanent (that is, lasting several years). Some domestic SW stations accidentally change frequencies when their transmitters "drift" off their assigned frequencies. Such conditions might last only a few days or may be a constant fact of life for a few stations, with their operating frequency changing slightly daily. Some domestic SW broadcasters might leave the air for months or years at a time and then suddenly return; the reason behind these disappearances is usually equipment failure and the time required to import spares to repair the equipment.

Domestic SW broadcasters may be government-operated, religious, or commercial. Not all government-operated domestic SW stations are located in the Third World; Australia, the USSR, and Canada all use shortwave to reach isolated areas of their vast territories. Religious broadcasters on domestic shortwave tend to have a strong service orientation toward their listeners, carrying programs on such topics as health care for an audience that's largely illiterate. Some of these stations are operated by evangelical Protestants, while others (particularly

FIGURE 7-1

A colorful souvenir of a memorable listening experience—a QSL from Office de Radiodiffusion Television de Senegal for reception on 4892 kHz. Its interval signal at sign on is a tuned played on the "cora," a stringed instrument indigenous to Senegal.

in Latin America) are operated by the Catholic Church. Commercial domestic SW stations are plentiful, especially in Latin America. Such stations are supported by advertising, and you'll hear ads for such international brands as Ford and Sony in addition to more local products. Sometimes you'll hear ads that are personal greetings to persons in distant towns and villages; this is particularly true in Latin America.

Tuning domestic SW stations is a challenge, and you might want to wait until you've gained experience with the easier international broadcasters before tackling them. But the rewards of listening to domestic SW stations are greater than the efforts necessary to hear them!

Europe

It might seem surprising that the highly advanced nations of western Europe, with their myriad AM, FM, and TV outlets, would support domestic SW broadcasting. Yet several domestic SW stations do operate in Europe and can be heard with varying degrees of difficulty in North America.

Domestic shortwave broadcasting is really lively in Italy. In the early 1980s, the state-run Radiotelevisione Italiana (RAI) lost its monopoly over broadcasting in Italy due to an unexpected court ruling. The result was that numerous private (and at first unlicensed) broadcasting stations were established on AM, FM, and SW. While the situation has stabilized somewhat with the introduction of formal licensing procedures and standards for private broadcasters, the SW broadcasting scene is still thoroughly Italian with all the delightful chaos that entails. Shortwave stations appear and vanish with astonishing speed, and you'll have to read a commercial SW magazine or join a SWL club to keep up with latest developments. Among those active at the time this was being written include the Voice of Europe on 7538 kHz, Radio Europe on 7295 kHz, and Radio Italia Internazionale on 7140 kHz. Most of these are heard in the 0600 to 0800 UTC time period. Perhaps the best heard is the Italian Radio Relay Service on 9815 kHz at their sign on around 0600 UTC. This station, like the other Italian private broadcasters, mainly relays programming prepared by others instead of originating its own.

Albania's major domestic network is relayed on both 5020 and 5057 kHz by Radio Gjirokaster. The only language you'll hear spoken is Albanian, often with alternating male and female announcers, along with much music with a distinctly local flavor. The audio quality of Radio Gjirokaster tends to be poor and some slight frequency variation can be expected. Your best chance of hearing this station is at sign on around 0400 UTC. Both frequencies use 50 KW but may not operate simultaneously; it's best to check both.

The changing situation in both East and West Germany will likely have an impact on the domestic shortwave broadcasting scene in both countries. Many of the West German domestic shortwave stations were probably more listened to in East Germany or by Germans traveling or living in the rest of Europe than in West Germany itself. At the time this book was being written, there was no way to tell how the reunification of Germany would affect these broadcasters and it's likely the situation could have changed greatly by the time you read this.

Perhaps the easiest German domestic station to receive in North America is Sudwestfunk, an AM and FM broadcasting network headquartered in Baden-Baden. They operate a transmitter on 7265 kHz relaying their normal programming. Unfortunately, this frequency is in the North America 40-meter amateur radio allocation, and interference can be heavy. The best time to try is after 0400 UTC on a weeknight to approximately 0600, when the sun rises in Germany. At this time, many hams are leaving the air to go to bed, and the frequency is more clear of QRM. The programming is all in German, with many German and English pop tunes, and you can often hear "Baden-Baden" in station identification announcements. A more challenging target is Suddeutscher Rundfunk on 6030 kHz, where it often suffers heavy QRM from international broadcasters. Again, the best time to try for this is after 0400 UTC or any time from your local sunset to about 0600 UTC if the frequency is clear of powerful international stations.

For years, East Germany's domestic radio services were relayed on shortwave by Stimme der DDR (Voice of the DDR), but in mid-1990 this was discontinued and replaced by a new service known as Deutschlandsender. At the time this was being written, it was scheduled at 2300 to 0530 UTC on 6115 kHz, although the QRM is also usually heavy on this frequency. The fate of this station after reunification is unclear.

USSR Domestic Shortwave

The Soviet Union is a massive nation stretching across two continents—Asia and Europe—with segments of its population living and working in extremely isolated areas. This makes shortwave the most practical method of broadcasting to such citizens, resulting in an extensive network of domestic SW services. The changes underway in the USSR will doubtlessly have some impact on domestic shortwave broadcasting there, but the extent and direction of these changes isn't clear as this is being written.

In the previous chapter, we saw how some of the various republics of the USSR have their own international broadcasting services. In the same manner, domestic SW broadcasting has a strong regional slant. There are several major national (or "all union," to use the Soviet term) broadcasting networks in the USSR. The largest is the program known as "Soyuz" (Russian for "union"), and is a combination of news, talk, and music. A second is called "Mayak," which means "lighthouse" in Russian. Mayak is primarily music, with news on the hour and half-hour; live sports events are sometimes broadcast. Mayak is broadcast continuously. (The others generally operate from approximately dawn to midnight, local time; the USSR stretches across eleven time zones.) Another national network is mainly classical music and literary, including readings of poetry and dramatic works.

The various republic broadcasting stations generally transmit two programs of their own. The first consists of "Soyuz" with news and various local features added at the station. The second program originates in the republic itself and is in the main language spoken in the republic. In addition, several regions and areas also have their own programming with more local items such as weather reports and activities; these are in the language mainly used in the particular area. Finally, major cities such as Moscow and Leningrad also produce their own local programs. The trend toward decentralization in the USSR will likely see more local networks and services instituted.

Reception of USSR domestic SW stations in North America centers around two periods. The first is from approximately 2000 to 0000 UTC, and the second begins around 1000 UTC to about your local sunrise. These reception times are due to the fact that many USSR domestic SW stations are found on frequencies below 6000 kHz. Unfortunately, trying to determine which station you're listening to can be difficult, as all identifications are in Russian or the language of the republic and the cities mentioned refer to the *studio* location rather than the transmitter site. Thus, the Russian words "Govorit Moscow" (This is Moscow) could be heard over a transmitter located somewhere in central Asia or Siberia. Moreover, available transmitter site data is hard to come by; the annual publication *Passport to World Band Radio* usually has the most accurate data currently available.

Among the most commonly heard USSR domestic SW frequencies are 4040, 4055, 4610, 4785 (in the Azerbaijan SSR), 4860, 4940, 4958 (also in the Azerbaijan SSR), and 5015 kHz. In addition, the Soyuz program is relayed on 49-meters and above by some of Radio Moscow's transmitters. There is a

Russian language equivalent to the Radio Moscow World Service known as Radiostantsiya Rodina (Radio Station Motherland), known simply as "Rodina" to SWLs. This is intended for Soviet citizens living abroad as well as the Soviet naval forces and merchant ship fleet traveling the world's oceans. This service relays segments of the Mayak program along with special features for seamen and Soviet citizens abroad. There are also special Russian regional services for seamen, such as Radiostantsiya Tikhiy Okean (Radio Station Pacific Ocean).

One major change in recent years has been the response of various Soviet regional stations to reception reports. For many years, the various Soviet regional stations simply ignored reception reports altogether. Now a growing number of regional stations do send QSLs in response to correct reports. Radiostantsiya Rodina now does so for most of its transmitter sites, as do several republic stations. This has resulted in several DXers now specializing in DXing and QSLing the Soviet regional and local outlets. As the USSR continues to change internally, the domestic SW broadcasting scene there will change with it and continue to be of interest.

Africa

Domestic SW broadcasting from Africa is really lively. There's a large number of stations you can hear and there's lot of variety. Most of the domestic stations here are run by national or regional governments, and some stations or frequencies may go silent for extended periods due to a shortage of funds to either operate the station or buy needed spare parts.

Perhaps the easiest place to start is South Africa, since much of the programming is in English and the transmitter facilities are first-rate. The easiest to hear in most parts of North America is Radio Five on 4880 kHz from 0400 UTC to a little past 0600. This is a commercial service with light pop music of the "easy listening" variety. Another that can be heard in the same time period is Radio Orion on 4810 kHz, another commercial service that plays a lot of instrumental music similar to what you often hear in elevators. There are also local services in the other languages of South Africa, such as Afrikaans. Radio Suid Afrika on 3320 kHz can be heard in North America from about 0515 UTC using Afrikaans. All of the South African domestic SW services are excellent verifiers.

South Africa formerly controlled Namibia when it was known as South-West Africa. While Namibia was under their control, the South Africans installed modern transmitting facilities which are still used by Radio Namibia. Programming here tends to be pop music, with many American artists, and multilingual station identifications on the hour and half-hour. The best bet seems to be 3290 kHz around 0200 UTC; there is also a station on 3270 kHz which can provide good reception when interference from utility stations is not too great.

At one time, Nigeria operated an extensive network of domestic SW stations which broadcast in English and local languages. While a lack of funds has put several stations off the air, Nigeria still has stations which can be heard reasonably well. Perhaps the best heard is the Radio Nigeria station at Kaduna on 4770 kHz; it can be heard after 0530 UTC in English with lots of ads and disco-style music. Another well-heard station is Radio Nigeria at Lagos on 4990 kHz around 2200 UTC to sign off after 2300 with English programming. A station with an interesting but tragic past is the Radio Nigeria outlet at Enugu on 6025 kHz. During the 1960s, Enugu was the capital of the breakaway nation of Biafra during Nigeria's bloody civil war (which was basically a tribal conflict between the dominant Ibo tribe of the would-be Biafran nation and the other tribes composing Nigeria). The station at Enugu was known as the Voice of Biafra, and allowed SWLs to literally hear a revolution in progress and get a brief glimpse into the problem of tribalism in Africa. Although the Voice of Biafra is now history, you can listen to its former facilities on 6025 kHz from about 2200 to sign off at 2300 UTC. You can also try after 0430 UTC, but QRM from major international broadcasters will usually be too severe.

Another African nation which uses plenty of English in its domestic SW broadcasting is Botswana. They also have one of the most unusual interval signals you'll ever hear—cow bells and the sound of farm animals. (SWLs worldwide were heartbroken when it turned out the interval signal was composed entirely of sound effects made by playful members of the station's technical staff!) Radio Botswana can best be heard at sign on at 0300 UTC on 3356, 4830, and 7255 kHz. For years, Radio Botswana was notorious for its refusal to QSL correct reports, but happily that situation has now changed.

A more challenging target is Radio Zambia International on 4910 kHz. At the time this book was being written, it was signing on around 0245 UTC with a striking bird call interval signal. It generally runs English until about 0415 UTC, when it switches to local languages. English news and commentary are given around 0345 to 0400 UTC most days, and these will let you see how events in Africa are seen from the perspective of a black-ruled nation.

Since France was a major colonial power in Africa, it's not surprising that you'll hear plenty of French on domestic SW stations in Africa. In countries where numerous local languages are spoken, French may often be the common language for government and business. One such country is Togo, whose national broadcasting service Radiodiffusion Television Togolaise, has a French name. It can be heard on 5047 kHz, where it operates a 100 KW transmitter, either at its 0530 UTC sign on or just before its 0000 UTC sign off. The frequency may vary somewhat, and most programs are in French. Listen for the French identification as "Ici Lome," which is the nation's capital and where the studios are located. Togo also operates a separate station at Kara, which can be heard on 3222 kHz before its 2300 UTC sign off or at its 0530 UTC sign on. Kara carries separate programming from the 5047 kHz station and identifies as "Ici radiodiffusion de Kara." In addition to French, you'll also hear some exotic local languages used by both stations.

A station that's easy to confuse with Togo is Radio Centrafrique, from the Central African Republic, on 5034 kHz. Like Togo, Radio Centrafrique uses a powerful transmitter but often drifts off its normal frequency. Listen carefully for its station identification of "Ici Radio Centrafrique" and its interval signal of a piano. It can be heard at its sign on of 0430 UTC and is often heard very well during winter in eastern North America before its 2300 UTC sign off.

FIGURE 7-2

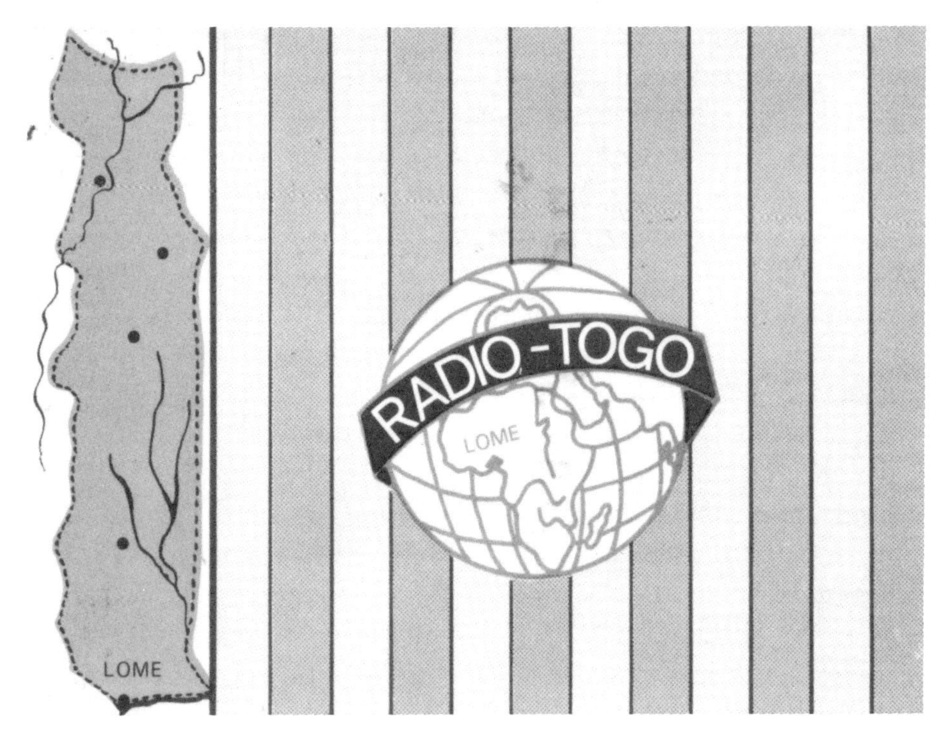

Togo's powerful signal on 5047 kHz is one of the first domestic shortwave stations most SWLs notice from Africa.

"Burkina Faso" may sound like the name of a character from a James Bond novel, but it's actually the new name of the nation formerly known as Upper Volta. Its broadcasting service, Radiodiffusion Television Burkina, can be heard on 4815 kHz at its 0530 UTC sign on and before its 0000 UTC sign off. You'll hear both French and "high life" music along with some lengthy talks.

Gabon is home to a commercial broadcaster known as Africa Number 1, which operates on numerous frequencies with powerful transmitters. Its income is derived primarily from serving as a relay base for such broadcasters as Radio Japan and Radio France Internationale. If you spend any time on the international broadcasting bands, you'll eventually run across Africa Number 1 or one of its relays. More interesting is Gabon's government-run domestic SW service, Radiodiffusion Television Gabonaise. They currently are found around 4777 kHz (although this frequency varies) at their 0630 UTC sign on or until their sign off at 0000 UTC. It can also be heard on 7220 kHz at sign on there around 0700 UTC. You're likely to hear "high life" and other pop music. Listen for station identifications beginning with "Ici Libreville," which is the nation's capital.

Ever heard a "tam-tam"? That's the interval signal used by La Voix de la Revolution in Benin on 4870 kHz. To hear a tam-tam, listen for a few minutes before their 0500 UTC sign on (this is weekdays only).You can also hear their second station at Parakou on 5025 kHz, which signs on at 0400 UTC. In addition to French, you'll hear some English used as well; African and western pop music is featured. Listen for the station identification of "Ici la voix de la revolution Beninoise."

In the northern part of Africa, Arabic replaces French as the language you're most likely to hear. However, some French influence remains present located in former French colonies and territories. In particular, French remains the language of business and government in these areas, and many SWLs have

FIGURE 7-3

Benin can often be heard well throughout North America on 4870 kHz.

had better luck with reception reports in French with these stations. One example is Algeria, whose Radiodiffusion Television Algerienne relays its domestic Arabic and French networks on shortwave. It can be heard in Arabic on 11715 kHz from 1800 to 2100 UTC and in French on 9640 kHz until sign off at 2200 UTC; sometimes there's a brief English segment around 1900 UTC.

Portugal also had colonial interests in Africa, and as a result Portuguese is often heard over domestic stations in their former colonies. One example is Radio Nacional de Angola, where Portuguese is the language of government and business but over ten different vernaculars are spoken in everyday use. 9535 kHz operates around the clock with most programs in Portuguese, and best reception is after 0500 UTC. Back in the mid-1970s, this station was exciting to listen to as Angola celebrated its newly won independence. I heard many seemingly live political rallies, complete with impassioned speeches, excited crowd shouts, and cheering. The programming is much calmer now, with music, sedate announcers, and even some telephone call-in shows.

The Near and Middle East

For years, this has been the world's premier "hot spot" for wars and near-wars. Unfortunately, you won't find these are good source of insights or information unless you speak Arabic, Farsi, or another language used in the region. However, you can enjoy the music and recitations from the Holy Quran despite the language barrier.

The domestic programming of the Broadcasting Service of the Kingdom of Saudi Arabia can be heard on 9720 kHz at 0300 UTC sign on and on 21505 kHz until 1800 UTC sign off. Special programs devoted exclusively to recitations from the Holy Quran can be heard on 11730 kHz from 0600 to 0800 UTC.

The United Arab Emirates (UAE) is home to two well-heard stations. The government-run Voice of the United Arab Emirates can be heard in English from 2200 to 0000 UTC and in Arabic from 0000 to 0200 UTC on 9595 and 11985 kHz. The English segment carries such programs as "The Truth About Islam," marking one of the few efforts made to propagate Islam via shortwave to non-Islamic audiences. It can also be heard entirely in Arabic from 0600 to 1600 UTC on 21515 kHz. The commercial UAE Radio and Television can be heard from about 0615 to 1630 UTC on 15435 kHz; most programs are in Arabic although there are some English segments.

Iran's Voice of the Islamic Republic can be heard on 15084 kHz from its sign on at 0730 UTC until its sign off at 0130 UTC. Languages used are Farsi and Persian, with much music and long, impassioned speeches. There is another outlet on 9022 kHz which has an English segment from 1930 to 2030 UTC; reception is best in eastern North America.

Iran's case shows that while Arabic is the language of Islam, not every Islamic nation has Arabic as its main language. Another is Turkey, whose principal language is (appropriately enough!) Turkish. Turkey is home to a major international

FIGURE 7-4

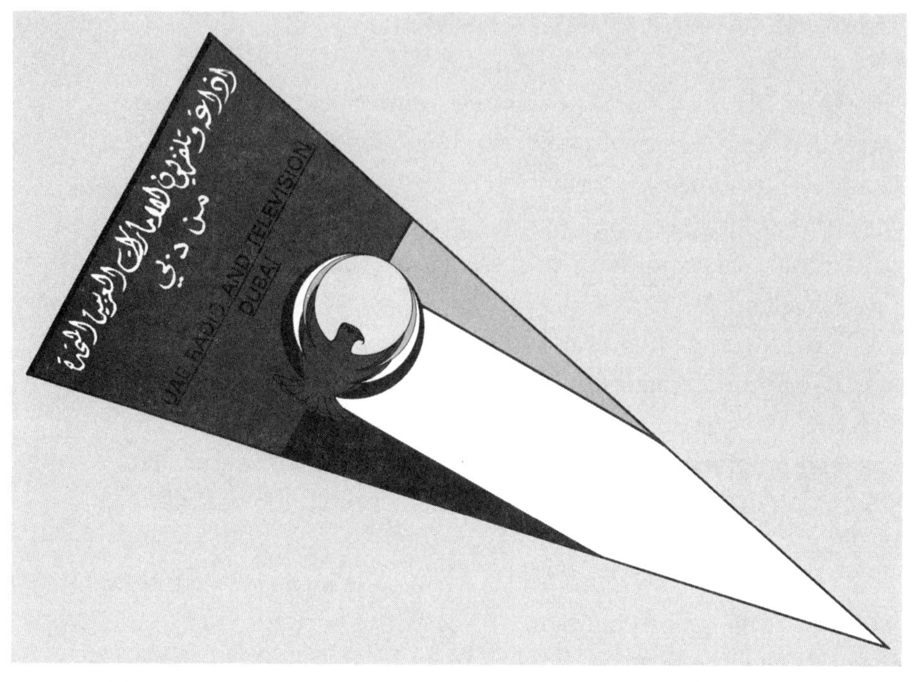

United Arab Emirates Radio and Television sometimes sends a pennant to listeners who report hearing their broadcasts.

broadcaster, the Voice of Turkey, which broadcasts in English to North America and is easy to hear. Much more challenging are two domestic SW stations which are exceptional catches in North America. One is Turkish Police Radio on 6340 kHz, which uses only 1000 watts of power. It is operated by the Turkish police to improve relations between police and citizens, and its programs consist of Turkish music and announcements. The best time to try for it is at its 0500 UTC sign on. The second station is operated by the Turkish State Meteorological Service on 6900 kHz with 2500 watts of power. Programming is music and, naturally enough, weather reports. It signs on at 0400 UTC, but QRM on this frequency from utility stations is usually severe.

The Indian Subcontinent

Indian domestic SW stations are invariably DX for North American listeners. The propagational path at any time or on any frequency is rough, and it takes patience and some help from the ionosphere to hear them. However, the exotic music and languages you can hear make the effort required worthwhile.

Broadcasting in India is the responsibility of the government's All India Radio. The domestic SW service is just as varied and complex as the nation itself. Over twenty different domestic SW stations use dozens of frequencies, and QSL collectors are pleased that most will verify reception reports. The programming could be in any of the languages spoken in India, such as Bengali, Hindi, Punjabi, Sanskrit, Tamil, Urdu, and English. The music will usually be the type described as "subcontinent music," which consists mainly of stringed instruments and flutes.

Most Indian domestic SW reception in North America is the result of gray-line propagation, meaning you should try for these stations a half hour before and after your local sunrise and sunset. Among the stations you might try for are Gauhati on 4775 kHz, Calcutta on 4820 kHz, Bombay on 4840 kHz, Delhi on 4860 kHz, Madras on 4920 kHz, and Aizawal on 5050 kHz. Your reception of any of these is likely to be only fleeting at best, lasting for just a few minutes.

Pakistan has a more modest domestic SW network, with only five stations. However, they may be even more challenging targets than the Indian stations. Outlets you might want to try your luck on include Karachi on 4815 kHz, Quetta on 4879 kHz, and Peshawar on 4950 kHz. The main language of Pakistan is Urdu, but you'll also hear some English. The music will be similar to what you can hear on All India Radio.

The Far East and Asia

The nations of Asia are diverse. They range from highly advanced, technology-based societies such as Japan to underdeveloped states locked into the past. Domestic SW broadcasters in Asia reflect this diversity.

Japan would seem at first to be an unlikely spot for domestic SW broadcasting, since it is heavily saturated with AM, FM, and TV stations. But Japan is the home of the Nihon Shortwave Broadcasting Company, which operates Radio Tanpa. This is a commercial broadcaster with all programs in Japanese, although some of the commercials have English phrases and many songs are in English. Reception of Radio Tanpa is more challenging in North America than Radio Japan, although it's not difficult. Two separate programs are transmitted by Radio Tanpa. The first operates on an almost 24-hour schedule on 3925, 6055, and 9595 kHz. The second operates from 2300 to 1300 UTC on 3945 and 6115 kHz. You'll probably find best reception on all frequencies beginning a couple of hours before your local sunrise.

Like the Soviet Union, China is a vast nation with many citizens located in isolated areas and uses shortwave to reach them. China's domestic SW stations are numerous and range from relatively easy SW targets to rare DX. For example, the stations at Shijiazhuang on 5860 kHz and Beijing on 5880 kHz are easy to hear in North America from 1100 UTC to your local sunrise. But just try to hear the station at Kunming, Yunnan on 2310 kHz at the same time!

Most domestic SW outlets in China are part of the Central People's Broadcasting Station (CPBS) network. There are also various regional stations and networks, and a separate service known as the Voice of the Strait intended for Taiwan. On such stations, you'll hear plenty of traditional Chinese music and opera along with a sprinkling of western culture (such as Linda Ronstadt). Reception of many stations is made easier by the

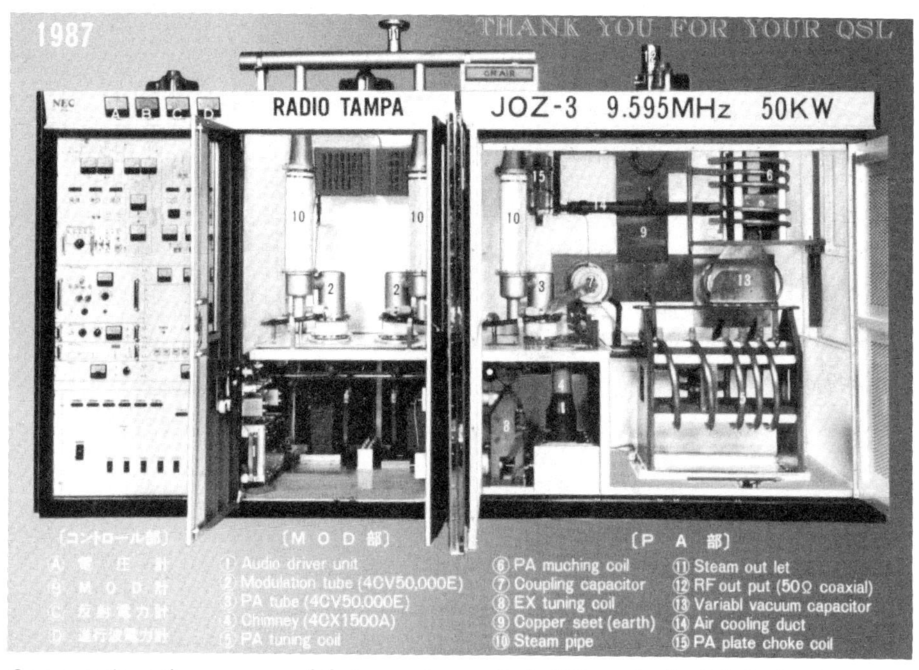

Some stations show scenes of their country on their QSL cards. Japan's Radio Tanpa lets you see what the inside of their transmitter looks like.

oddball frequencies they use. Among the ones in use as this book was being written were 3815, 4100, 4250, 4330, 4460, 4500, 4750, 5163, 5320, 6400, 6750, 6840, and 6937 kHz. If you tune these channels from about 1100 UTC to your local sunrise, you'll probably hear most of these stations if conditions are normal. A good reference such as the current edition of *Passport to World Band Radio* is a big help in determining which location and service you're listening to.

Vietnam is a nation that still looms large in the American consciousness. Its external service, the Voice of Vietnam, was widely quoted in the American news media and was often the first source of information about positions taken by the Hanoi government. (The American press also managed to continuously misidentify this station as "Radio Hanoi"—a name the

FIGURE 7-6

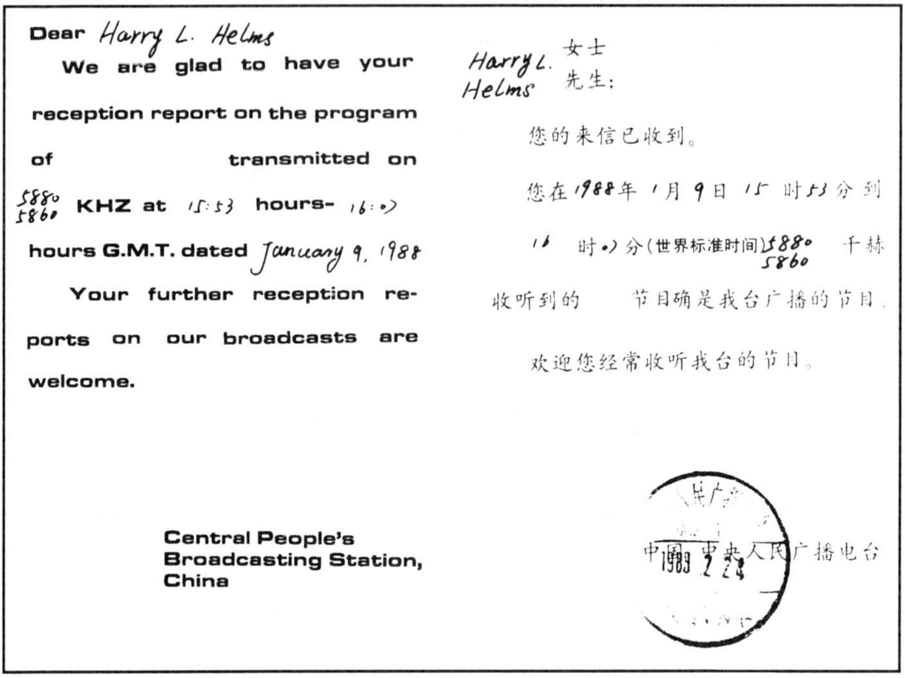

Dear Harry L. Helms

We are glad to have your reception report on the program of transmitted on 5880 5860 KHZ at 15:53 hours- 16:02 hours G.M.T. dated January 9, 1988

Your further reception reports on our broadcasts are welcome.

Central People's
Broadcasting Station,
China

Harry L. Helms 女士
先生：

您的来信已收到。

您在1988年1月9日15时53分到16时02分（世界标准时间）5880 5860千赫收听到的 节目确是我台广播的节目，

欢迎您经常收听我台的节目。

中国中央人民广播电台

An English and Chinese QSL card from the Central People's Broadcasting Station network in China.

station never used, leaving alert SWLs wondering which station, if any, the various news reporters were actually listening to!) The Voice of Vietnam is still in operation today, and Vietnam also operates an extensive domestic SW broadcasting network as well. The main station at Hanoi operates on 6450 and 10060 kHz and can often be heard well until approximately 1600 UTC. In addition, there are some local SW stations, such as Cao Bang on 6575 kHz and Lai Chau on 6641 kHz, which are prime DX targets in North America. All programming is in Vietnamese or Montagnard (a language spoken by a mountain people within the borders of Vietnam who have never accepted Vietnamese rule). All frequencies used by Vietnamese stations are highly variable.

Malaysia is a federation made up of a peninsula and part of an island. It's not surprising that domestic SW is used as part of the government's Radio Television Malaysia broadcasting system. The major languages of Malaysia are Malay, Chinese, and English, and you'll hear all three used by Radio Malaysia. The basic Malay service is broadcast continuously on 5965 kHz, and should be audible from about 1000 UTC to approximately an hour after your local sunrise if the frequency is free of QRM. The English home service was scheduled, at the time this book was being written, from about 0915 to 1600 UTC. Malaysia is divided into districts, and two have their own broadcasting systems. The Sabah district operates Radio Malaysia Kota Kinabalu on 4970 kHz, where it can be heard in Malay until 1600 UTC. The Sarawak district operates Radio Malaysia Sarawak on 4950 kHz, where its English and Chinese programs can be heard until about 1600 UTC. Since Malaysia has a large Moslem population, don't be too surprised if you hear recitations from the Holy Quran over one of these stations.

Singapore was once a part of Malaysia. It's now an independent state and a rising economic power. The Singapore Broadcasting Corporation's English network, Radio One, can be heard on 5052 and 11940 kHz, with best reception usually around your local dawn. Don't expect much in the way of exotic Oriental music; on these frequencies, the programming tends to be western pop music and talk shows.

Indonesia is composed of the numerous islands making up the Indonesian archipelago. Indonesia may well be the favorite target of DXers worldwide, as it has an extensive network over 70 different domestic SW stations operated by the government's Radio Republik Indonesia (RRI). Most of these stations operate in the tropical broadcasting bands with relatively low power, making them exceptional DX targets. The RRI stations tend to be excellent verifiers of reception reports, often with long personal letters in Indonesian. (In case your Indonesian isn't as fluent as you'd like, relax; several form letters and reporting guides for Indonesian have been developed and are available.) But the DX challenge and readiness to issue QSLs does not explain the popularity of Indonesian stations among DXers. Part of the reason may lie in the exotic nature of the programming itself. Indonesian is a truly unusual language to western ears, and many SWLs find a lilting, almost musical quality to it. RRI stations also play plenty of authentic "South Seas" music. RRI stations use a distinctive interval signal known as "Song of the Coconut Islands," played using an organ, piano, flute, and vibes, which is played just before the hour. On the hour are time "pips" followed by the news in Indonesian. Listening to domestic SW stations from Indonesia lets you get a taste of a culture that's vastly different from anything you've probably known before, but one with a powerful appeal!

Indonesian DXing, including positively identifying the stations you do hear, is an advanced listening activity that takes time and patience to master. Perhaps the easiest station to hear is RRI at Ujung Pandang, which operates on either 4719 or 4753 kHz (but not both simultaneously); best reception is around your local dawn. If you manage to hear Ujung Pandang, you can also try for Dili on 3307 kHz, Banjarmasin on 3250 kHz, Ternate on 3345 kHz, Tanjungkarang on 3395 kHz, Jakarta on 4775 kHz, and Ambon on 4845 kHz. Your best chance for hearing these stations is around your local dawn. Don't be too surprised if you find yourself becoming one of those DXers hooked on Indonesia!

Australia and the Pacific

Broadcasting in Australia is conducted by private commercial interests as well as the Australian Broadcasting Corporation (ABC). The ABC is responsible for three SW stations intended for reception in the nation's vast and sparsely inhabited interior. All programs are relays of the various ABC regional networks, which are broadcast on AM and FM as well as SW. Thus, all programs you hear will be in English (albeit the Australian version of it). The identification announcement is usually a simple "This is the ABC." And all stations will usually verify correct reception reports.

If you're an SWL in eastern North America, the ABC station at Perth may well be the most physically distant station you can hear on shortwave. The transmitter is located at Wanneroo, near Perth, on Australia's west coast and can often be heard with good signals throughout North America on 6140 and 9610 kHz from a couple of hours before your local sunrise until about 1600 UTC. During late evenings and at night during periods of high MUFs, this station can also be heard on 15425 kHz. A similar ABC station at Brisbane, on Australia's east coast, can also be heard in the same time period on 4920 and 9660 kHz.

Three other ABC stations are far more difficult to hear: Alice Springs on 2310 kHz, Tennant Creek on 2325 kHz, and Katherine on 2485 kHz. While all use 50 KW transmitters, the difficult propagation at those frequencies make them rare catches in eastern North America.

The nation of Papua New Guinea is a former Australian territory. While English is the common language of government and business, other languages such as Melanesian, Pidgin, and Hiri Motu, are spoken. Broadcasting is the responsibility of the National Broadcasting Commission of Papua New Guinea. Since the nation is large and relatively undeveloped, shortwave plays a major role in linking the nation together. Currently, there are several domestic SW stations in operation there. A few are relatively easy to hear in North America, but most are difficult. The easiest one to try for is the National Service outlet at Port Moresby on 4890 kHz, which can usually be heard until its scheduled sign off at 1300 UTC. Programs here will be in English, Pidgin, and Hiri Motu, and the station will verify correct reports. Other Papua New Guinea stations, such as Radio West New Britain on 3235 kHz and Radio New Ireland on 3905 kHz, are much more difficult.

FIGURE 7-7

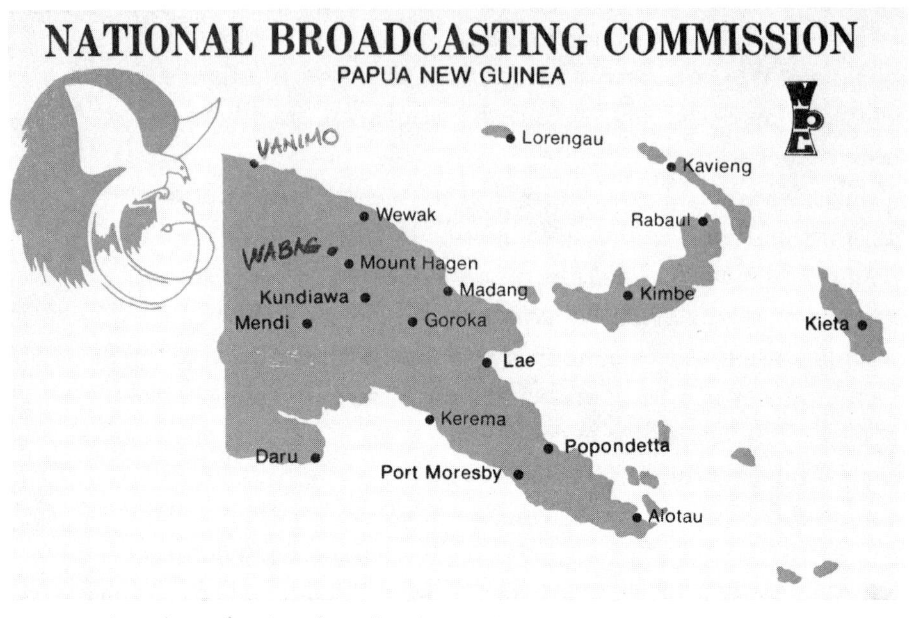

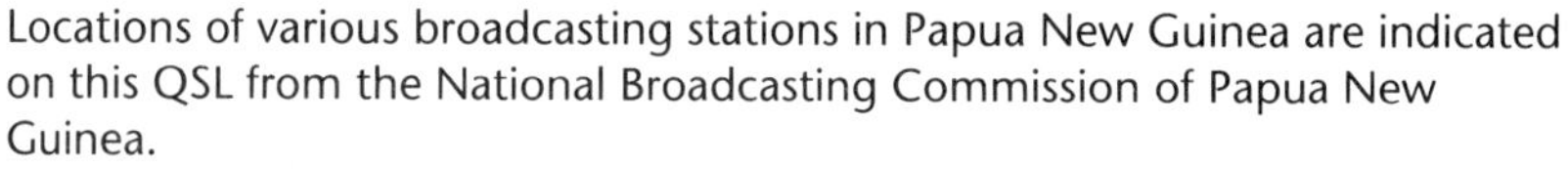
Locations of various broadcasting stations in Papua New Guinea are indicated on this QSL from the National Broadcasting Commission of Papua New Guinea.

Ask people to name an island paradise in the Pacific, and they're likely to mention Tahiti. Tahiti is a part of French Polynesia, and the principal languages spoken are French and Tahitian. Broadcasting there is conducted by the Societe Nationale de Radio Television Francaise d'Outre Mer (usually simply abbreviated RFO), a French governmental body. Fortunately for SWLs, they can usually be well heard in North America on 11825 and 15170 kHz after 0100 UTC. French is normally used until about 0300 UTC, with Tahitian thereafter until 0830 UTC. During the Tahitian segments, you can hear some of the traditional music associated with a tropical paradise, such as flutes, guitars, and choral voices. During the French programs, however, you'll hear French pop tunes and even disco music (proving, perhaps, that paradise isn't what it used to be). While RFO-Tahiti is currently reluctant to verify reports, in the past they sent out a QSL card depicting a topless mermaid.

A plain—but effective—QSL from the Solomon Islands Broadcasting Corporation. (What's an "effective" QSL? One that verifies your reception!)

Another French possession in the Pacific is New Caledonia. This has been the scene of recent conflicts between residents who wish to remain affiliated with France and those who want the island to become an independent state. RFO is likewise responsible for broadcasting in New Caledonia, and is most often heard on 7170 kHz from about 0700 UTC until your local sunrise. Most programming is in French, and music is mainly French pops.

The Solomon Islands are located just east of Papua New Guinea, and their principal languages are Pidgin and English. The government operates the Solomon Islands Broadcasting Corporation, which often identifies itself in English as "Radio Happy Isles." Programs alternate between English and Pidgin segments, and can often be heard with good signals on 9545 kHz until 0800 UTC or on 5020 kHz until 1200 UTC.

North America

Canada, like Japan, seems an unlikely spot for domestic SW broadcasting. However, the northern regions of Canada are some of the most rugged and isolated areas anywhere, and many people living there depend on SW radio for their links with the outside world.

The Canadian Broadcasting Corporation operates two shortwave stations on 6160 kHz that are superb DX targets. CKZN, in St. John's, Newfoundland, operates with only 300 watts of power and relays AM station CBN while CKZU, in Vancouver, British Columbia, uses 500 watts to relay AM station CBU.

FIGURE 7-9

Toronto's CFRB on 1010 kHz is relayed on shortwave by CFRX on 6070 kHz, enabling listeners throughout North America to get the latest Toronto traffic reports direct from the CFRB airplane.

Both of these broadcast in English. CKZN is currently scheduled from 0930 to 0500 UTC while CKZU operates from 1500 to 0900 UTC. Since both stations operate on the same frequency, it's easy to confuse the two—provided you can hear them at all! As you might expect, the biggest problem here will be QRM from international shortwave broadcasters.

There are also five private commercial AM stations in Canada that relay their signals on shortwave. The easiest of the group to hear is CFRX, 6070 kHz, in Toronto, which relays AM station CFRB. Try for this one whenever there's a propagation path between you and Toronto (including midday if you're within a few hundred miles of Toronto). The transmitter power is 1000 watts, but that's enough to put a good signal into most of North America. The second easiest target is probably CHNX in Halifax, Nova Scotia, on 6130 kHz. This relays station CHNS and is often heard in eastern North America shortly before dawn when interference on the frequency is usually at its lowest level. CHNX uses 500 watts. Another station often heard in eastern North America around the same time is CFCX in Montreal on 6005 khz. It's another 500 watt station relaying CFCF.

The other Canadian domestic SW broadcasters are much more formidable DX challenges. They are CFVP on 6030 kHz in Calgary, Alberta (using 100 watts) and CKFX in Vancouver on 6080 kHz. The transmitter power of the latter is a whopping ten watts! But don't give up hope—it's been heard in eastern North America when the frequency is clear.

Latin America

For the purposes of this section, we'll define "Latin America" as all of Central and South America plus the Caribbean region. This definition isn't accurate from a geographical or cultural perspective, but does make sense for radio propagation and reception.

Not everything "south of the border" will be in Spanish. For example, Radio Belize can sometimes be heard on 3285 kHz until about 0600 UTC with mostly English programs. The Voice of Guyana can be found around 5950 kHz in English at its 0730 UTC sign on or at its 0200 UTC sign off. At the time this book was being written, both stations operated irregularly on SW due to equipment problems. However, when they're on, they provide a fascinating insight to very different English-speaking cultures; you can hear local announcements of birthdays and funerals, a "prayer of the day," an ad for "Tropical Motors," and a rap tune—all in the same fifteen minute period!

English also shows up on some of the evangelical domestic SW broadcasters. One example is HRVC, La Voz Evangelica, in Tegucigalpa, Honduras on 4820 kHz. As you might surmise from its name, the station is an evangelical Christian station owned by the Conservative Baptist Home Mission Society in Kansas. During the 0300 to 0500 UTC period on weeknights, this station broadcasts English religious programs; most are the same ones heard on AM and FM religious broadcasters in the United States. Other stations, such as Costa Rica's Radio Limon on 59545 kHz, also have brief English segments.

French is also heard from Latin America. In addition to being the location of a Radio France Internationale relay base, it also has an RFO station serving the local population. It can be heard on 5055 kHz in French and operates continuously.

Languages such as English and French are definitely in the minority, though. The vast majority of stations will be in Spanish, with Portuguese (from Brazil) a distant second. You'll also find a few stations with programs in local Indian languages such as Quechua, but Spanish is easily dominant.

Fortunately, Spanish is an easy language for English speakers to deal with. Many words and phrases are similar to their English equivalents. With a little practice, you'll be able to recognize station names, identification announcements, and even some commercials. Although almost all Latin American stations have been assigned call letters, don't expect to hear them used on the air. Instead, most Latin American stations use slogans such as "Radio Los Andes" or "Radio Cristal" to identify themselves. This works in your favor if you don't speak Spanish, since the slogan or name is much easier to catch than individual letters of the alphabet.

One thing you'll quickly notice is that there's no such thing as "Latin American music." The music is just as varied as the nations making up the continent. Marimba music will indicate you're listening to a station in Guatemala or a neighboring country. You'll run across music best described as "sad flutes"; this is heard on stations in Peru, Ecuador, and other nations through which the Andes pass. With a little practice, you'll be able to associate certain types of music with certain countries.

The obvious place to start in any examination of Latin America is Mexico. One surprising aspect of Mexican domestic SW broadcasting is how few stations such a relatively large nation has; many smaller Latin American countries have far more. Part of this is due to the number of AM stations there, with many using high power; there are few areas of Mexico where several AM stations can't be heard easily. Another factor may be influence of American broadcasting and receiver manufacturing patterns. Since so much broadcasting and receiving equipment in the formative days of Mexican broadcasting was imported from the United States, it's not too surprising that broadcasting and listening habits followed the American model.

None of the Mexican domestic SW broadcasting stations are easy to receive unless you live near the southern border of the United States. Among those you can try for are XEOI, Radio Mil, on 6008 kHz from 1600 to 2200 UTC, XEQ, Radio XEQQ, on 9680 kHz from 1200 to 0600 UTC, XEUDS, Radio Universidad de Sonora, 6115 kHz, from 1300 to 0500 UTC, and XEQM, Su Pantera, on 6105 kHz from 1200 to 0500 UTC.

All domestic Mexican SW stations tend to operate irregularly. A station may be silent for months (or even years) and then abruptly return to the air.

Cuba has a well-known international service, Radio Havana Cuba, but only one domestic SW station is located there. Radio Rebelde is a broadcasting network in Cuba, and is relayed on SW on 5025 kHz. The operating schedule is 1030 to 0400 UTC, although this station also operates irregularly.

Guatemala is relatively easy to hear thanks to TGNA, Radio Cultural, in Guatemala City on 3300 and 5955 kHz. This is another evangelical Christian station, and English is often heard from 0300 to 0430 UTC. Other Guatemalan stations you can hear include La Voz de Nahuala on 3360 kHz and Radio Tezulutlan on 3370 kHz. Best reception of these stations is at their 1100 UTC sign on.

Honduras is easy to hear via the previously mentioned HRVC, La Voz Evangelica, on 4820 kHz. A more challenging target to try for is another religious broadcaster, Radio Luz y Vida, in Santa Luis on 3250 kHz. This can be heard until its 0400 UTC sign off most winter evenings; there is sometimes some English after 0300 UTC on weekends.

Costa Rica has several targets to try for, including Radio Reloj on 4832 kHz. This station operates all night and is easily heard throughout North America. Another relatively easy station is Faro del Caribe on 5055 kHz, which often has English around 0300 to 0400 UTC.

The Dominican Republic shares the island of Hispaniola with Haiti. It is home to one of the best known domestic SW broadcasters, HIUA, Radio Clarin, in Santo Domingo. During the late 1970s, this station broadcast an English segment known as "This is Santo Domingo," hosted by Rudy Espinal. This quickly became a cult favorite among SWLs, primarily due to Rudy's personality and approach to the program. In the 1980s, some American SWLs began producing a program titled "Radio Earth" which Radio Clarin broadcast. Rudy Espinal was again host of this program, but the broadcasts ended as the Radio Earth organization made plans for their own broadcasting facility. By the mid-1980s, Radio Clarin was relaying programs produced by various anti-Castro groups. While but a shadow of its former self at the time this book was being written, Radio Clarin might have more surprises for SWLs in the future. It is currently scheduled at 1100 to 0200 UTC on 11700 kHz; 9955 kHz is also used at times.

The two most often heard countries from Latin America are Colombia and Venezuela. At night in North America, the 60-meter tropical band seems packed with stations from these two countries until they begin to sign off around 0500 or 0600 UTC. (A growing number of stations now operate all night.) Among the stations from Colombia you can hear are Caracol

FIGURE 7-10

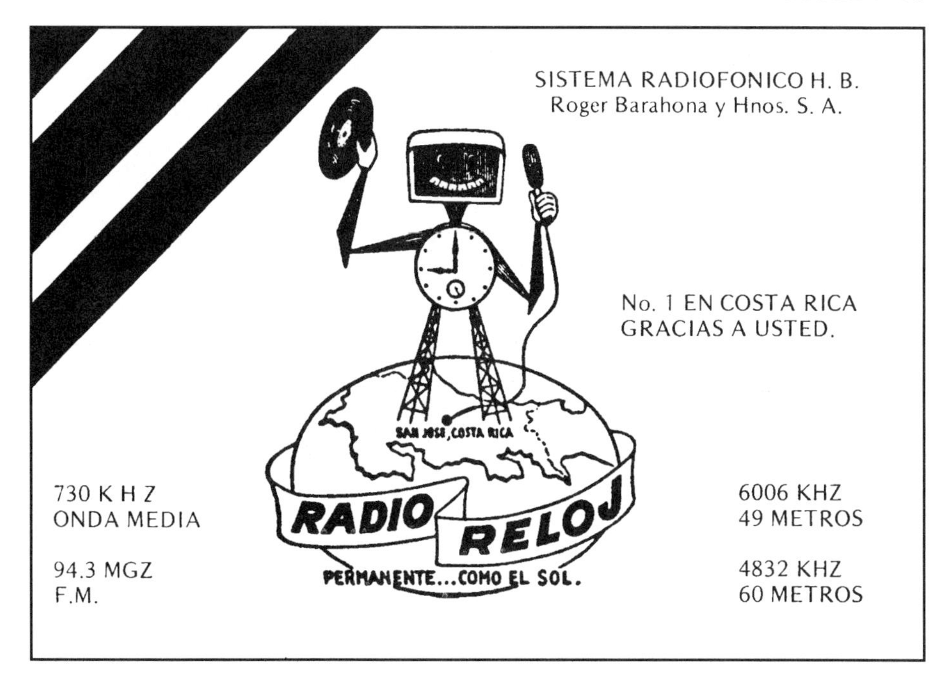

Radio Reloj's (pronounced "ray-low") booming signals from Costa Rica on 4832 kHz are easy to hear.

Bogota on 4755 kHz, La Voz de Cinaruco on 4865 kHz, Ondas del Meta on 4885 kHz, La Voz de Rio Arauca on 4895 kHz, Caracol Neiva on 4946, Radio Florencia on 4976 kHz, and Caracol Bogota on 5075 kHz. Venezuelan stations commonly heard include La Voz de Carabobo on 4780 kHz, Radio Tachira on 4830 kHz, Radio Valera on 4840 kHz, Radio Capital on 4850 kHz, Radio Continental on 4940 kHz, Radio Rumbos on 4970 kHz, and Ecos del Torbes on 4980 kHz. Colombia and Venezuela also have several domestic stations active on other SW bands. But if you're a new SWL the action from those two countries on 60-meters alone will keep you busy!

Ecuador is easy to hear via HCJB. More authentically Ecuadorean are the numerous SW broadcasters found there. Those widely heard include Radio Popular on 4800 kHz, Radio Quito on 4920 kHz, and Radio Catolica on 5030 kHz. These can be heard in the evening hours until about 0500 UTC.

Brazil breaks the pattern in Latin American broadcasting in several ways. As mentioned earlier, the prime language here is Portuguese instead of Spanish. While Brazil has numerous stations on the tropical broadcasting bands, its domestic broadcasters also make wide use of the higher frequency ranges. One of the best bands for reception of Brazilian stations in North America is the 25-meter international broadcasting band. Among the stations that can be heard in the evening hours in North America are Radio Globo in Rio de Janeiro on 11805 kHz, Radio Bandeirantes in Sao Paulo on 11925 kHz, and Radio Nacional Amazonia in Brasilia on 11780 kHz. Another well-heard station is Radio Inconfidencia in Belo Horizonte on 15190 kHz. At the other end of the spectrum, several Brazilian stations on the 120-meter and 90-meter tropical broadcasting bands rank with the toughest DX catches of all.

Peru is more difficult to hear than Brazil, but many of its stations are easily spotted since they operate on unusual frequencies outside the allocated broadcasting bands. An example is Radio Nor Andina in Celendin on 4462 kHz, which can be heard around its 0900 UTC sign on. Another is Radio Cutervo on 6691 kHz until its 0200 UTC sign off. Perhaps the best-heard station is Radio Atlantida on 4790 kHz until its 0400 UTC sign off.

Bolivia can be a surprisingly difficult country to hear. Many of its domestic SW stations are low-powered or are covered by QRM from stronger stations in Colombia or Venezuela. Perhaps the best heard is Radio Panamerica in La Paz on 6105 kHz at their 1100 UTC sign on or in the evenings until their 0400 UTC sign off.

Reception of Paraguay can be tough even for experienced DXers. However, the government's Radio Nacional in Asuncion can be heard on 9735 kHz with a good signal most evenings until its 0400 UTC sign off.

Uruguay is tough. Perhaps the best bet is Radio El Espectador in Montevideo on 11835 kHz around 1030 UTC or in the evenings until its 0400 UTC sign off.

Chile is relatively easy via the government's Radio Nacional, in Santiago, on 15140 kHz during late afternoons and early evenings in North America. Sometimes English is used here for news and station identifications. A more difficult target is Radio Santa Maria on 6030 kHz from 0900 to 0300 UTC.

Argentina has an international broadcasting service known as Radiodiffusion Argentina al Exterior, a holdover from the days of Juan Peron. All domestic SW stations in Argentina are government-owned at the time this book was written. The main domestic SW service, known as Radio Nacional, can be heard throughout the night hours in North America on 6060 kHz until sign off at 1100 UTC.

This quick tour of domestic shortwave broadcasting has only scratched the surface of what you can hear. Don't get discouraged at the more difficult reception or strange languages—your efforts here will be more than rewarded!

CHAPTER EIGHT

Utility Stations

AS MENTIONED IN THE FIRST CHAPTER, utility stations "do work" and are not intended for reception by the general public. That doesn't keep many SWLs and DXers from avidly tuning in; that's hardly surprising, since most of the stations you can hear on SW are utility stations. Utility station listening is anything but routine—you might find yourself listening to a Presidential telephone call from Air Force One or coded instructions to Strategic Air Command bombers aloft.

Unlike the other types of stations mentioned in this book, there are a couple of legal restrictions on listening to utility stations. The Electronic Communications Privacy Act (ECPA) of 1986 was intended mainly to protect the privacy of cellular telephone calls, and is generally not applicable when listening to frequencies below 30 MHz. However, it does prohibit listening to "remote pickup" units used by radio stations for live remote broadcasts for shopping centers, car dealerships, high school football games, etc. These units are found in the 25 to 26 MHz range and use FM. How do you know if you're listening to one of these prohibited remote pickup stations? Basically, you can't. And, moreover, there's no practical way you (or anyone else) could possibly be prosecuted under the ECPA for listening to such signals in the privacy of your home unless you confessed to a prosecuting attorney with time on his or her hands. The ECPA was little more than a public relations effort by the cellular telephone industry to convince their customers that cellular telephones offer the same degree of privacy that wired telephones do. A technically ignorant U. S. Congress passed the ECPA. They were fooled by it. You don't have to be.

A more important restriction, based upon international treaty, prohibits repeating the contents of anything you hear transmitted by a utility station (with the exception of a *marker* transmission, which will be discussed later) to anyone other than the station itself or intended recipient. This means that you shouldn't repeat *what* you heard over a utility station to anyone else. You can disclose the time and frequency you heard a utility station, its call sign (and the call sign of any stations it contacted), the mode of transmission used, and a general idea of what the transmission involved but not its specifics. For example, it's okay to mention that you heard a utility station sending telegrams but not the contents of any telegrams, to whom they were sent, or who sent them. These secrecy statutes are backed by a federal law making violations subject to a $10,000 fine and a year in jail.

In practice, these prohibitions have no effect. I know of no one who's ever been prosecuted for violation of this particular law. In fact, the trend in recent years has been toward a rather flagrant and open disregard for it. For example, virtually all major news organizations listened in on Coast Guard and Navy frequencies during the search for the wreckage of the Space Shuttle *Challenger*. In fact, the first word that the remains of the astronauts had been located came not from NASA but from monitoring of the search ship's radio communications. The three major broadcast news networks, along with Cable News Networks, did more than merely paraphrase the contents of such transmissions; they rebroadcast them as received. If the "big boys" can get away with it, so can you.

You'll hear every mode mentioned in chapter 2 used by utility stations, although FM is mainly restricted to frequencies above

25 MHz. AM is seldom used except for standard time and frequency stations. RTTY and SSB are the most commonly used modes you'll hear. CW is still used widely, but is becoming less so each passing year. In fact, an international agreement to move most maritime communications to satellites may eliminate many of the CW stations currently in operation.

Longwave Beacons and Other Stations

"Longwave" is usually used to refer to all frequencies below the lower end of the standard broadcast band at 540 kHz. For many years, this was a sort of "no man's land" as far as SWLs were concerned since few had the necessary equipment to tune longwave. Now, most new "shortwave" radios tune down to about 150 kHz, and longwave listening is undergoing a major growth surge.

Reception on longwave is a lot like listening on the standard broadcast band. During the day, reception is usually limited to less than two hundred miles (with greater ranges possible over water paths such as oceans), with night reception covering distances of several hundreds or even thousands of miles. DX reception is better in the winter than summer, with the equinoxes being particularly good times.

One major problem with longwave reception is electrical noise. The random noise generated by power lines, neon signs, motors, and other electrical devices is strongest at longwave frequencies, and your location might be too noisy for good longwave reception. To combat noise, some listeners use a loop antenna (as mentioned in chapter 4) to "null out" sources of electrical noise. Other listeners use portable receivers to listen away from noise sources.

Since most receivers start tuning at 150 kHz, let's start looking at longwave from that point. The first interesting range is 160 to 190 kHz, which is used by some types of wireless intercoms. It is also used by some experimenters, as FCC rules permit operation of unlicensed transmitters there as long as transmitter power is less than about one watt. Most of these experimental stations operate in CW, usually as beacons that repeat either the initials of the operator or the QTH of the station.

From 190 to 405 kHz, you'll hear stations which seem to do nothing but repeat one to three characters in Morse code continuously—and that Morse code is transmitted in AM using audio tones. You'll also find some stations giving weather forecasts with Morse code in the background. These stations are known as *longwave beacons*. Since loop antennas can be used for direction-finding, these beacons can be used for aeronautical and maritime navigation. Those beacons which include weather information usually do so at 15 and 45 minutes after the hour, although sometimes this is more frequent.

Call signs of beacons don't follow international allocations. Instead, they usually suggest the name of the town or airport at which the beacon is located. Table 8-1 is a sampling of beacons; note the calls. Most suggest the location, but the origins of some beacon calls are obscure.

Trying to figure out where a beacon is located can be a problem. Beacons normally don't include their location, and beacons change call sign or frequency—or leave or go on the air—with bewildering rapidity. The best reference work available is the *Aero/Marine Beacon Guide* by Ken Stryker and Joe Woodlock. For keeping current on latest information about new, changed, or deleted beacons, membership in the Longwave Club of America is a good investment.

From 405 to 512 kHz, you'll find a few more beacons and several CW stations used for two-way maritime CW communications. These stations will be in "true" CW (that is, you'll need your receiver's BFO to copy the signals) and you'll find the communications will involve coastal stations (which use three-letter call signs) and shipboard stations (which normally use four-letter call signs).

TABLE 8-1 Typical Longwave Beacons

kHz	Call sign and location	kHz	Call sign and location
194	TUK, Nantucket, MA	371	ITU, Great Falls, MT
206	GB, Galveston Bay, TX	375	TGE, Guatemala City, Guatemala
223	DM, Detroit, MI	380	BBD, Brady, TX
236	GNI, Grand Isle, LA	393	FBG, Fort Bragg, NC
281	RSZ, Tempe, AZ	408	HBD, Youngstown, OH
305	RO, Roswell, NM	410	DAO, Fort Huachuca, AZ
315	USR, Simon Reyes, Cuba	414	PCW, Port Clinton, OH
322	S, Point Sur, California	415	CBC, Cayman Islands
323	BSD, St. David's Head, Bermuda	417	HHG, Huntington, IN
326	BHF, Freeport, Bahamas	430	LML, Loma Linda, Colombia
329	CH, Charleston, SC	434	WLO, Mobile, AL
340	BDG, Blanding, UT	478	CLA, Havana, Cuba
344	JA, Jacksonville, FL	484	KLC, Galveston, TX
353	FME, Fort Meade, MD	516	VPX, Pineville, WV
363	RNB, Millville, NJ	521	GM, Greenville, SC

It's here that you'll first notice marker transmissions. A marker is the radio equivalent of someone at the microphone saying "Testing one, two, three." A marker transmission is transmitted by a coastal station repeatedly to allow ship stations to tune it in, and to "hold" a frequency between communications. Markers on CW usually follow these two forms:

CQ CQ CQ DE UFL UFL UFL K
VVV VVV VVV DE WNU WNU WNU K

"CQ" is an old radio signal which is a general call to anyone who might be listening; it is an invitation for any receiving station to call the transmitting station (in this example, UFL in Vladivostok, USSR). "DE" is French for "from," and is followed by the station's call sign. The "K" is the radiotelegraph equivalent of "over," and signals the end of a transmission. Some markers begin "VVV." This has no meaning, and simply indicates a test transmission. You might also find "QRA" and

"QSX" used in markers. In such a context, "QRA" asks for the call signs of any stations that might be listening and want to make contact, while "QSX" indicates the station is listening for replies.

Markers are easy to recognize. The code speed is usually slower than normal CW communications, and the repeated sequence of characters stands out. You'll soon learn to recognize the sound of CQ ("dahdidahdit dahdahdidah"), VVV ("dididi-dah didididah didididah"), and DE ("dahdidit dit') in Morse code. With the Morse table in the appendix and a tape recorder, you can replay the marker until you have it completely decoded. (If you do this long enough, you'll learn the Morse code without meaning to.)

500 kHz is an international emergency frequency for ship stations. 512 kHz is a *calling* frequency. A calling frequency is one on which stations make contact with each other and then switch to another (or *working*) frequency.

The frequencies from 512 to 540 kHz are filled with another assortment of beacons similar to those found from 190 to 405 kHz, except you won't hear any weather broadcasts over these. 530 kHz is often used for low-powered stations giving road and airport information. These can sometimes be heard at a surprising distance.

Maritime Communications

Hundreds of frequencies are used for maritime radio. In fact, more stations are probably active in the maritime service than in any other type of shortwave utility.

Most SSB maritime transmissions will be in USB and will be *duplex*. This means the ship station and the coastal stations it contacts will use separate frequencies instead of the same frequency. By international agreement, such frequency pairs have been standardized and assigned reference numbers. As you tune through the maritime shortwave bands, you'll run across numerous "one-sided" communications; these mean you're listening to one half of a duplex pair.

There are also some *simplex* frequencies, on which both stations use the same frequency and take turns transmitting and receiving in a formal fashion. Table 8-2 lists some of the more active simplex frequencies, and what you can hear on them.

Two frequencies that can provide much interesting listening at night if you're within a few hundred miles of a major body of water are 2182 and 2670 kHz. 2182 kHz is the international distress and calling frequency. You won't have to listen long here before hearing a call from a vessel in trouble. Usually, the problem is relatively minor, such as a disabled engine, but sometimes true life-and-death emergencies (such as a sinking ship) can be heard. This frequency is monitored continuously

TABLE 8-2 Important SSB Marine Radio Frequencies

kHz	Use
2182	International distress and calling
2670	U. S. Coast Guard weather broadcasts
4125	Ship calling and intercoastal waterway traffic
4143.6	Ship and coastal stations
4376	U. S. Coast Guard traffic
4419.4	Coast stations
6218.6	Ship stations
6221.6	Ship stations
6518.8	North American inland and intercoastal ships
8257	Ship calling
8291.1	Ship and coastal stations
8294.2	North American inland and intercoastal ships
8718.9	U. S. Coast Guard traffic
8765.4	U. S. Coast Guard traffic
12429.2	U. S. Coast Guard traffic; civilian ships and coastal stations
12435.4	Ship and coastal stations
16523.4	Ship stations
16587.1	U. S. Coast Guard traffic; civilian ships and stations
22124	Ship and coastal stations

by the U.S. Coast Guard and other services, as well as by other ships at sea. The 2670 kHz channel is used by the U.S. Coast Guard for its marine weather broadcasts, and stations along the Atlantic and Pacific coasts transmit these bulletins several times each hour. The Coast Guard also uses 2670 kHz for routine communications between its vessels and shore bases. In fact, 2670 kHz is almost never quiet at night; something always seems to be going on there.

The remaining frequencies in table 8-2 are filled with a variety of merchant vessel and coastal station traffic. Some of the more interesting listening on these channels comes during unusual or difficult weather situations; a ship caught in a squall or hurricane provides dramatic listening.

Much maritime communications are still carried on by CW, and all maritime bands above 4000 kHz have substantial CW segments humming with activity. The activity you'll hear is very similar to that heard on the longwave marine segment, with standard call signs used for coastal and ship stations and plenty of marker transmissions. Table 8-3 gives the call signs and locations for some of the more commonly heard CW maritime coastal stations. You'll also hear many of these same stations on RTTY, which now carries a major chunk of messages that used to go via CW.

As noted earlier, though, time is running out for many SW maritime stations. The increasing use of satellite technology aboard ships threatens to relegate SW communications to a secondary role. This is already taking place in the United States. In recent years, well-known and well-heard coastal stations such as WSL in Amagansett, New York and WOE in Lantana, Florida closed because of a declining volume of SW communications. Others will doubtlessly close as well, and the current volume of CW activity on the maritime bands could be only a memory within a few years.

Many maritime stations will verify SWL reports, and some even have their own printed QSL cards. In other cases, station

TABLE 8-3 CW Maritime Station Call Signs and Locations

Call	Location
AME3	Madrid, Spain (Spanish Navy)
CCS	Santiage, Chile (Chilean Navy)
CFH	Halifax, Nova Scotia, Canada
CLA	Havana, Cuba
CTU	Monsanto, Portugal (Portuguese Navy)
CUL	Lisbon, Portugal
C6N	Nassau, Bahamas
DAF	Norddeich, Germany
DZU	Manila, Philippines
EAD	Aranjuez, Spain
EBC	Cadiz, Spain (Spanish Navy)
FFL	St. Lys, France
FUE	Brest, France (French Navy)
FUF	Fort de France, Martinique (French Navy)
GKE	Portishead, Great Britain
GYC	London, Great Britain (British Navy)
GYQ	Portsmouth, Great Britain (British Navy)
GYU	Gibraltar (British Navy)
HEB	Berne, Switzerland
HKC	Buenaventura, Colombia
HLG	Seoul, South Korea
HPN	Balboa, Panama
HWN	Houilles, France (French Navy)
IAR	Rome, Italy
ICB	Genoa, Italy
KLB	Seattle, WA
KLC	Galveston, TX
LFW	Rogaland, Norway
LSA	Boca, Argentina
LSO	Buenos Aires, Argentina
LZW	Varna, Bulgaria
JCK	Kobe, Japan
JCT	Chosi, Japan
JOS	Nagasaki, Japan
MTI	Plymouth, Great Britain (British Navy)
NAM	Norfolk, VA (U. S. Navy)
NMC	San Francisco, CA (U. S. Coast Guard)
NMF	Boston, MA (U. S. Coast Guard)
NMO	Honolulu, HI (U. S. Coast Guard)
NMR	San Juan, Puerto Rico (U. S. Coast Guard)
NPG	San Francisco, CA (U. S. Navy)
NPN	Agana, Guam (U. S. Navy)
OMC	Bratislava, Czechoslovakia
OST	Ostende, Belgium
OXZ	Lyngby, Denmark
PCH	Scheveningen, Netherlands
PJC	Curacao, Netherlands Antilles
PPR	Rio de Janeiro, Brazil
PWB	Balem, Brazil
PWZ	Rio de Janeiro, Brazil (Brazilian Navy)
RIH	Khiva, Uzbek SSR, USSR (Soviet Navy)
ROT	Moscow, USSR (Soviet Navy)
SAB	Goteberg, Sweden
SPH	Gydnia, Poland
SVD	Athens, Greece
SVF	Athens, Greece
TIM	Limon, Costa Rica
VHR	Darwin, Australian (Australian Navy)
VRT	Hamilton, Bermuda
UAT	Moscow, USSR
UDH	Riga, Latvian SSR, USSR
UFL	Vladivostok, USSR
UJE	Moscow, USSR (Soviet Navy)
UJY	Kaliningrad, USSR
UMV	Murmansk, USSR
URD	Leningrad, USSR
UXN	Arkhangelsk, USSR
VIP	Perth, Australia
VWB	Bombay, India
WCC	Chatham, MA
WLO	Mobile, AL
XFE	Ensenada, Mexico
XFM	Manzanillo, Mexico
XSG	Shanghai, China
ZLP	Irirangi, New Zealand
ZSC	Cape Town, South Africa
4XZ	Haifa, Israel
6WW	Dakar, Senegal (French Navy)
6YI	Kingston, Jamaica
8PO	Barbados
9YL	Trinidad

personnel will sign and return a prepared card sent by a SWL with the report. Other stations will not verify under any circumstances or even acknowledge letters from listeners. Addresses for utility stations and details of their verification policies can be found in various SWL publications and club bulletins.

Fixed Stations

This is one category where DX opportunities have markedly declined in recent years. Two decades ago, there were numerous fixed stations used for overseas telephone calls. Such stations signed on with a voice marker transmission that went something like "This is a test transmission for circuit adjustment purposes from a station of the American Telephone and Telegraph Company. This station is located near New York City." Such stations often verified reception reports, and QSLs from these outlets are collector items today. Only a few of these stations remain, as satellites have taken over the bulk of international telephone traffic.

Today, most fixed stations use RTTY. Some fixed stations do use SSB and CW, however, and these stations are listed in table 8-4. Several of these are backup or emergency communications networks supporting satellite systems. While these are not used for normal communications, they are often tested on a regular basis for proper operation.

International organizations such as the International Police Organization (Interpol) and the International Red Cross use shortwave for some of their communications. Contrary to popular image, Interpol itself has no agents and makes no arrests; its primary function is as a clearinghouse for information about criminal activity. Much Interpol traffic is in RTTY, although 10390 kHz is widely used for CW transmissions. The International Red Cross commonly uses 20753 kHz for USB communications related to its humanitarian functions.

TABLE 8-4 A Sampling of Fixed Stations

kHz	Call, location, and other data
5133	KAA60, Federal Communications Commission (FCC), Grand Island, NE, and other FCC stations nationwide (RTTY)
5167.5	State of Alaska emergency frequency, several stations
7474.9	KCP63, Federal Aviation Administration (FAA), Longmont, CO, and other FAA stations nationwide (LSB)
9311.5	KEC96, Federal Bureau of Investigation (FBI), New York, NY, along with other FBI stations across the country (USB)
10390	FSB57, International Police Organization (Interpol) headquarters, Paris, France, plus several other Interpol stations worldwide (CW)
10493	WGY900, Federal Emergency Management Agency (FEMA), Washington, DC, along with other FEMA stations nationwide (USB)
16077	U. S. Army Corps of Engineers national operating frequency in USB
20753	HBC88, International Red Cross Headquarters, Geneva, Switzerland, and several other Red Cross stations worldwide (CW, USB, and LSB)

You'll also notice what seem to be international broadcasting stations using SSB in the fixed bands. These are *feeder* stations, used by some international broadcasters to relay programs from their studios to transmitting sites. These stations are gradually being phased out as satellites link the sites.

Time and Frequency Stations

WWV, WWVH, and CHU are not the only standard time and frequency stations you can hear. Table 8-5 lists some of those that have been heard in North America. Naturally, European stations are better heard along the east coast, while the Asian stations are better heard along the west coast. Like WWV/WWVH and CHU, other time and frequency stations transmit time pulses each second; some also include propagation forecasts and voice announcements as well. Many of these stations

TABLE 8-5 Standard Time and Frequency Stations

kHz	Call and location	kHz	Call and location
2500	WWV, Fort Collins, CO	10000	BPM, Xian, China
2500	WWVH, Kauai, Hawaii	10000	JJY, Tokyo, Japan
3330	CHU, Ottawa, Ontario, Canada	10000	LOL, Buenos Aires, Argentina
3810	HD210A, Guayaquil, Ecuador	10000	WWV, Fort Collins, CO
4996	RWM, Moscow, USSR	10000	WWVH. Kauai, Hawaii
5000	IAM, Rome, Italy	10004	RID, Irkutsk, USSR
5000	IBF, Turin, Italy	14670	CHU, Ottawa, Ontario, Canada
5000	LOL, Buenos Aires, Argentina	14996	RWM, Moscow, USSR
5000	WWV, Fort Collins, CO	15000	BPM, Xian, China
5000	WWVH, Kauai, Hawaii	15000	JJY, Tokyo, Japan
5000	ZUO, Olifantsfontein, South Africa	15000	LOL, Buenos Aires, Argentina
		15000	WWV, Fort Collins, CO
7335	CHU, Ottawa, Ontario, Canada	15000	WWVH, Kauai, Hawaii
7600	HD20A, Guayaquil, Ecuador	15004	RID, Irkutsk, USSR
8000	JJY, Tokyo, Japan	20000	WWV, Fort Collins, CO
9996	RWM, Moscow, USSR		

verify reception reports; figure 8-1 shows a QSL card received from JJY in Tokyo.

While WWV/WWVH and CHU identify at regular intervals, such is not the case with other time and frequency stations. Some only identify in Morse code while a few never identify; all you'll ever hear are pulses each second from these stations. And not all stations operate continuously—a few operate as little as five minutes at a time. The latest operating schedules can be found in a current edition of *Passport to World Band Radio* or the *World Radio TV Handbook*.

The best time to hear stations other than WWV or WWVH on 5, 10, or 15 MHz is around your local sunrise and sunset, when the effects of changing ionospheric conditions can disrupt reception of those two stations enough to permit other stations to slip through under them.

Standard time and frequency station JJY in Tokyo sends out this colorful QSL card for correct reports of its signals.

Aeronautical Communications

USB is extensively used for aeronautical work between aircraft flying international routes and airports. And language isn't a major obstacle, since English is the language of international aviation. The only other major languages you're likely to hear used are French (mainly for flights into and out of Africa) and some scattered Russian (for flights to and from the Soviet Union).

"VOLMET" is an acronym formed from the French words for "flying weather," and are aviation weather broadcasts giving information on weather conditions enroute and at various destinations. Table 8-6 gives the frequencies and originating

TABLE 8-7 International Civil Aviation Frequencies

kHz	Use
3016	North Atlantic flights
5550	Caribbean flights
5568	Central American flights
5574	Eastern Pacific flights
5598	North Atlantic flights
5616	North Atlantic flights
5667	Middle Eastern flights
6577	Caribbean flights
8825	North Atlantic flights
8843	Eastern Pacific flights
8846	Caribbean flights
8864	North Atlantic flights
8918	Caribbean flights, Middle Eastern flights
10017	Middle Eastern flights
11396	Caribbean flights
13291	North Atlantic flights
13297	Caribbean flights
13306	North Atlantic flights
13312	Middle Eastern flights
13354	Eastern Pacific flights
17907	Caribbean flights
17946	North Atlantic flights

airports for selected VOLMET transmissions. In most cases, there is more than one VOLMET station on a frequency. In such cases, the stations take turns transmitting. Identification will usually be "This is New York Radio," "This is Shannon Aeradio," or a similar form. Note the frequencies are regional and serve flights along different aeronautical routes.

Several frequencies have been set aside for SSB communications by flights and airports along various international routes. Transmissions on these frequencies are essential communications for the safety of the flight, such as arrival times, fuel consumption, weather information, and the like. These frequencies are listed in table 8-7. Several other frequencies are reserved for long distance operations control (LDOC) use. These are noncritical, "internal" communications between an airline and its aircraft aloft. Table 8-8 lists some frequencies and the major airlines that use them.

Some of the most interesting aeronautical listening involves the branches of the United States military. Table 8-9 lists some of the more widely used frequencies and the services which use them; all transmissions will be in USB.

One thing you'll quickly notice is how fond the different services, particularly the Air Force, are of so-called "tactical" call signs. You'll hear stations identify themselves as "Elastic,"

TABLE 8-8 In-Flight LDOC Channels

kHz	User(s)
4654	Swissair
5532	KLM, Alitalia
6637	Air France, El Al, JAL, Lufthansa, Quantas
8921	British Airways
8924	Aeroflot, El Al, KLM, Varig
10030	Aeroflot
10072	British Airways
10075	Middle Eastern Airlines
11351	Air France

"Classical," "Morphine," "Rim Control," and similar, meaningless terms. Such call signs change on a daily basis. If they leave you confused, that's the entire point; any potential enemies are supposedly equally confused and are unable to identify the stations or "enter" the communications network without being detected.

Yet some "normal" identifications are used, and it is usually possible to at least determine which armed service you are listening to. The general Air Force identification format consists of a letter from the international phonetic alphabet, such as "Delta," followed by two or three digits. Air Force bases and ground stations often refer to themselves simply by their location or name, such as "Andrews" or "Vandenberg." Navy stations follow a similar format, although ground stations are usually preceded by the word "Raspberry."

The Civil Air Patrol (CAP) is a civilian organization sponsored by the Air Force, and is often active for search and rescue missions and emergency drills. Stations in CAP are assigned call signs from normal FCC allocations consisting of three letters followed by three digits. However, on-air activity is usually identified by a tactical call sign consisting of a combination of four letters and digits.

The Tactical Air Command (TAC) is responsible for the

Table 8-9 U. S. Military Aviation Frequencies

kHz	Use
4495	U. S. Air Force
4700	U. S. Navy
4725	U. S. Air Force/Strategic Air Command
4732	Air Force One
4737	U. S. Navy
5020	U. S. Air Force
5297	North American Aerospace Defense Command (NORAD)
5692	U. S. Coast Guard/U. S. Navy
5700	U. S. Air Force/Strategic Air Command
6670	U. S. Air Force
6697	U. S. Navy
6701	U. S. Navy
6720	U. S. Navy
6730	U. S. Air Force/Tactical Air Command/Air Force One
6750	Air Force One
6756	Air Force One
6761	U. S. Air Force/Strategic Air Command
6833	U. S. Navy
8101	U. S. Air Force/Strategic Air Command
9002	U. S. Navy
9014	U. S. Air Force/Tactical Air Command
9018	Air Force One
9023	U. S. Air Force/Strategic Air Command
9027	U. S. Air Force/Strategic Air Command
9036	U. S. Navy
9057	U. S. Air Force
11118	Air Force One
11191	U. S. Navy
11243	U. S. Air Force/Strategic Air Command
11258	U. S. Navy
13201	U. S. Air Force/Air Force One
13204	Air Force One
13231	U. S. Navy
14905	U. S. Air Force/Civil Air Patrol
15041	U. S. Air Force/Strategic Air Command
15091	U. S. Air Force/Strategic Air Command
17982	U. S. Navy
18027	Air Force One
20631	U. S. Air Force/Strategic Air Command
23337	U. S. Air Force/Strategic Air Command
26617	Civil Air Patrol
26620	Civil Air Patrol

defense of North America against enemy aircraft. The various TAC ground stations identify themselves by the word "Raymond" or "Fireside" followed by one or two digits.

Those frequencies identified in table 8-9 as belonging to the Strategic Air Command (SAC) will be of particular interest. SAC is responsible for nuclear missiles and bombers, and all of the frequencies in table 8-9 are part of its "Giant Talk" communications network. Don't expect to understand any of the messages you might hear on an SAC frequency; as you would expect, all are carefully encoded. Often you may hear a transmission on an SAC frequency begin with the word "Skyking." This is generally believed to be a general message to all SAC nuclear forces. These messages are usually quite short, and consist of numbers and letters from the phonetic alphabet. What they refer to, and what they mean, are unknown and beyond the ability of a casual listener (and, one hopes, any hostile or potentially hostile nation) to determine.

Listening to military communications has developed into something of a specialty within the SWLing hobby. The best way to keep up to date on happenings in the field is to join a SWL club with good utility coverage. *Monitoring Times* magazine has particularly good coverage of military communications.

Radioteletype Communications

The introduction of microcomputer-based radioteletype (RTTY) terminal devices has resulted in an explosion of interest in RTTY monitoring. It's now the fastest growing segment of utility listening.

RTTY reception is so different from the other types of listening covered in this book that it's impossible to do more here than just cover the highlights. Fortunately, several good books on RTTY are available from SWL equipment suppliers and book dealers and one of them is essential if you get interested in RTTY listening (or "reading" is perhaps more like it!).

As mentioned back in chapter two, RTTY is a system of sending characters via shifting a radio signal between two different frequencies. "RTTY" itself covers a variety of different methods of sending characters by shifting a signal's frequency. Perhaps the most common is *Baudot*, the earliest system. This permits sending of all ten digits, common punctuation marks, and all letters of the alphabet as capital letters. Another system, *ASCII*, permits transmission of small and capital letters of the alphabet in addition to numbers and punctuation. (ASCII is the same method used for storing and transmission of characters in personal computer systems.) More elaborate systems, such as the *forward error correction* (FEC) and *answer request* (ARQ) modes, include error-checking functions. In two-way communications, these modes can prompt the sending station to retransmit a section of text that was received garbled or with characters missing.

What can you "see" on RTTY? Much of it will be similar to that on the maritime CW bands, with several markers from various stations. You'll also see long strings of "RYRYRYRYRYRYRYRYRY" being sent as well. This is a tuning signal for RTTY stations; a receiver is tuned so that the "RY" is received clearly. You'll also see various "test sentences" containing every letter of the alphabet similar to those used by typists. When material is transmitted, a lot of it will be text of news dispatches in various languages. TASS, the Middle East News Agency, Agence France Presse, the U. S. Information Agency, and other news services use RTTY to send news to client newspapers and organizations. Other RTTY communications are equivalent to the type of traffic related to shipping and operations passed over CW maritime stations. And some RTTY traffic will consist of coded weather information in the form of number blocks.

Since the military forces of the world are heavy users of RTTY, it's no surprise that a lot of RTTY traffic will be encoded. When received, such signals will print as random gibberish and make no sense whatsoever. However, not all gibberish output will necessarily be encoded traffic. Foreign languages which do not use the Roman alphabet, such as Japanese, Arabic, and Russian, will also "print funny" when received. Unlike true encoded traffic, however, the output they produce will not be random; patterns will repeat for various words in these languages and the groups will be of varying length.

To get an idea of what you can currently receive on RTTY, check the RTTY column in *Popular Communications* magazine. RTTY is a bit difficult to get involved in as your first SWLing activity, but is something more and more SWLs are trying every day.

CHAPTER NINE

Other Radio Activities

AS MENTIONED BEFORE, "shortwave listening" involves more than just listening to stations transmitting from 1600 to 30000 kHz. In this chapter, we'll look at some "nonshortwave" SWL activities, some of which involve *transmitting* as well as receiving.

Broadcast Band DXing

Many SWLs (like me) first experienced reception of distant radio signals while tuning the standard AM broadcast band (abbreviated BCB), which ranges from 540 to 1600 kHz. Some SWLs stick with it and make DXing the BCB their prime listening interest. A few top BCB DXers have managed to hear over 100 different countries entirely between 540 to 1600 kHz.

It's easy to get started in BCB DXing. Tune across the BCB anytime from between your local sunset and sunrise, and you'll hear several stations located hundreds or even thousands of miles away. Many of these stations verify reception reports, and with a little effort and time you can probably collect QSLs from stations in 25 different states as well as from neighboring countries such as Canada, Cuba, and Mexico. In fact, any ordinary radio tuning the BCB should be adequate to let you hear about half the states, several Canadian provinces, and a few foreign countries.

But once you reach those totals, you'll find yourself hitting a wall. You'll notice that the same stations tend to be heard on the same frequencies each night, with only minor daily variations in what can be heard. To increase your totals of states and countries heard, you'll find that specialized equipment, antennas, and knowledge are needed. Geography also plays a crucial role in what you can hear. A BCB DXer located near the shore in a lightly populated area of the east coast will *always* be able to hear more countries than a DXer with similar skills and the same equipment located in the center of North America. In fact, the best locations for a BCB DXer in terms of reception possibilities are on islands, such as Hawaii, in the middle of the ocean away from major land masses.

One reason BCB DX is so challenging was discussed in chapter 5. The ionosphere always takes a bigger "bite" out of a signal at BCB frequencies than it does for signals at higher frequencies. Each refraction of a BCB signal weakens it more than a comparable refraction of a shortwave signal. Multihop reception is common on SW frequencies, but reception involving more than a couple of "bounces" is rare on the BCB. When it does happen, signals are subject to long periods of fading; you may find that only one minute of audio can be heard in a five minute reception period. Simply put, the ionosphere is usually kind to signals at SW frequencies but is murder on BCB frequencies.

Noise is another problem. Lightning from thunderstorms produces radio noise (or "QRN"—remember?). This noise, which is a radio signal, propagates well at BCB frequencies. It's not unusual for QRN from thunderstorms hundreds of miles distant to be heard—a listener in Chicago in the dead of winter can experience QRN from thunderstorms in the Gulf of Mexico. DXing the BCB often means a case of "ringing" ears from QRN; a good noise blanker in your receiver is a must for serious work.

The biggest problem, however, is the sheer number of stations found on the BCB. In North and South America, the BCB is divided into 107 "channels" spaced every 10 kHz beginning at 540 kHz and extending to 1600 kHz. In the United States alone, over 5000 stations are crammed into this range! A few of these stations only operate in the daytime; most others reduce their power and/or use directional antennas at night to avoid causing QRM to other stations on the frequency. Despite these measures, you'll find most channels will have several stations operating on them at night. Some form of directional antenna, such as a rotatable indoor loop, is a major help in sorting out these stations. (BCB DXers aren't the only ones frustrated by this congestion. Ordinary listeners have been abandoning the AM broadcast band in favor of FM at a rapid rate, and a key reason cited in listener surveys is the poor reception, particularly at night.)

FIGURE 9-1

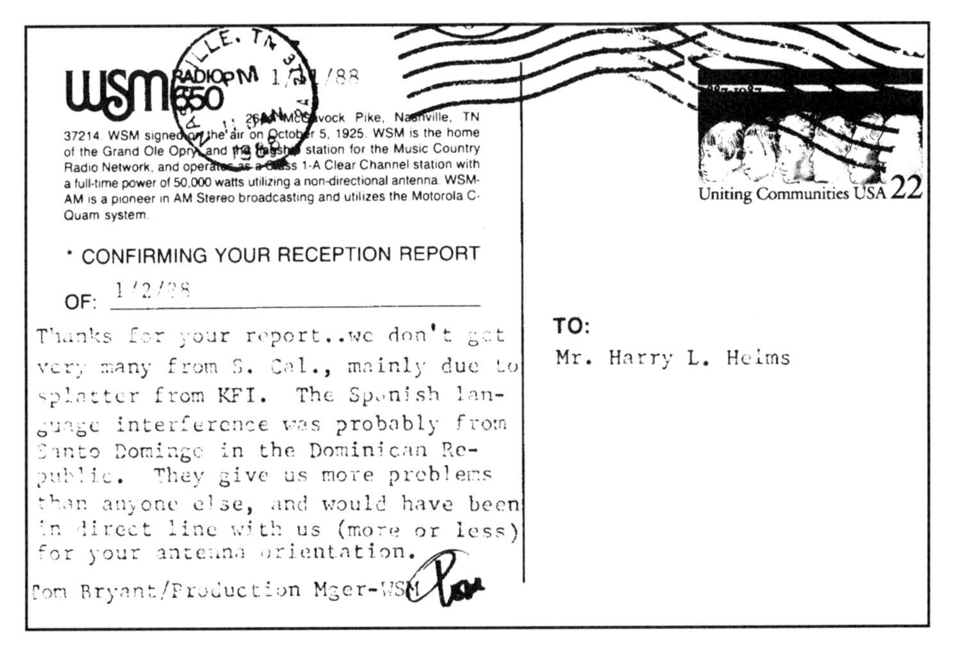

wsm RADIO 650

1/[illegible]/88

26[illegible] McGavock Pike, Nashville, TN 37214. WSM signed on the air on October 5, 1925. WSM is the home of the Grand Ole Opry and the flagship station for the Music Country Radio Network, and operates as a Class 1-A Clear Channel station with a full-time power of 50,000 watts utilizing a non-directional antenna. WSM-AM is a pioneer in AM Stereo broadcasting and utilizes the Motorola C-Quam system.

* CONFIRMING YOUR RECEPTION REPORT

OF: 1/2/88

Thanks for your report..we don't get very many from S. Cal., mainly due to splatter from KFI. The Spanish language interference was probably from Santo Domingo in the Dominican Republic. They give us more problems than anyone else, and would have been in direct line with us (more or less) for your antenna orientation.

Tom Bryant/Production Mger-WSM

Uniting Communities USA 22

TO:

Mr. Harry L. Helms

Although WSM, 650 kHz, in Nashville, Tennessee is supposedly a "clear channel" station, this QSL card from its shows that it suffers from QRM in much of the country.

Not all BCB stations are on 10 kHz spacing, however. Some stations in the Caribbean and South America operate on what are known as "split" frequencies, such as 655 and 1555 kHz, between the 10 kHz channels. Sometimes this is a deliberate choice by the station to avoid QRM, while in other cases it is the result of improper transmitter operation. Outside North and South America, BCB channels are spaced every 9 kHz rather than 10 kHz. This means that distant BCB stations in Europe, Africa, and Australia can sometimes "sneak through" between 10 kHz channels in North America. Since such stations may operate only one kHz or two from a loud North American BCB station, you'll need a receiver with superb selectivity to have any realistic chance at these stations.

In BCB DX terminology, a "domestic" station is considered to be any AM station in Canada as well as the United States. Domestic BCB stations fall into three categories: *clear channel*, *regional*, and *local*. This latter classification is sometimes referred to as the "graveyard" frequencies. These are 1230, 1240, 1340, 1400, 1450, and 1490 kHz; stations in the United States on these frequencies can use only 1000 W of power into a nondirectional antenna. Each channel has close to 200 stations operating in the United States alone, with the result being tremendous QRM. Spend a few minutes listening on one of these channels where you don't have a local station; it will probably sound something like the "rumble" in a concert or theater before the performance begins. DXing these frequencies is a test of your endurance; you can spend over an hour on a frequency waiting for one station to surface above the jumble long enough for it to be positively identified.

The term "clear channel" is misleading these days. In the past, it was used for stations which were the only ones on a given frequency. Such stations could be heard throughout most of the United States and Canada at night thanks to their 50 KW transmitters and nondirectional antennas. Decades ago, ordinary listeners throughout the country listened to such stations as WLW in Cincinnati and WSM in Nashville each night, much as today's viewers watch "superstations" like Atlanta's WTBS. Over the years, most clear channel monopolies were broken up by placing additional stations on clear channel frequencies and mandating the use of directional antennas. This practice divided the nation up into "service areas" for two or three "clear channel" stations per frequency, with minimal interference between stations.

The end of the clear channel concept began in the early 1980s. For years, many stations had been authorized to operate

FIGURE 9-2

QSL

OMAHA, NEBRASKA

KFAB

50,000 WATTS — AM 1110 KCS.

115,000 WATTS ERP — FM 99.9 MCS.

This will confirm your AM reception report of .... Thanks

KFAB in Omaha, Nebraska shares 1110 kHz with KRLA in Pasadena, California and WBT in Charlotte, North Carolina. All three use highly directional antennas to avoid causing interference to each other.

on clear channels during daytime; at sunset, they had to leave the air. Beginning in 1980, several of these stations were authorized to continue operations at night with directional antennas and/or reduced power. In addition, new stations were allowed to begin operations on clear channels. The result was a virtual end to the former wide coverage enjoyed by most clear channel outlets.

Regional stations are in the middle ground between clear channels and graveyards. These stations operate with powers of about 10 KW at night (some use powers as high as 50 KW in the daytime) and all use directional antennas. Many regional stations use different antennas for day and night operations.

Most domestic AM stations now broadcast continuously, although some stations sign off around their local midnight. A handful of stations are still "daytimers," operating from local sunrise to sunset. The pattern of sunrise/sunset changes—power levels, antenna patterns, stations returning to or leaving the air—make sunrise and sunset favorite listening times for domestic DX. Listening at such times means you have to be alert. A station may abruptly change its power or antenna pattern, leaving a second station "in the clear." This situation may last only a few minutes, until the second station also adjusts its power or antenna pattern, leaving a third station present on the frequency. Those few minutes could well be your only chance to hear the second station.

BCB stations sometimes go off the air for maintenance or to conduct tests after their local midnight. If a station leaves the air during this time, it could open up the channel and let you hear new stations. Sometimes, stations will test using their daytime power and/or antenna pattern, and this gives you a chance to hear stations that you otherwise can't. Monday mornings are usually the best time to catch such tests, although they can be heard any night of the week.

While the chances for domestic BCB DX have been declining, there are still opportunities to hear foreign stations.

Since so many foreign BCB stations operate on split frequencies, receiver selectivity is important; a receiver equipped with a narrow bandwidth mechanical or crystal filter will be an enormous help. In many situations, some form of exalted carrier SSB reception (as described in chapter 3) must be used. Some form of directional antenna, such as a loop, is also important.

The first foreign BCB stations you're likely to hear will be from the Caribbean, Central America, and South America. Some of these, such as Radio Cayman on 1555 kHz, operate on split frequencies. Others operate on standard 10 kHz channels, but are audible when ionospheric storms disrupt reception of domestic stations on the same frequencies.

Hearing stations from the other side of the Atlantic or Pacific depends a lot on your location. If you're along the east coast, you'll have a good chance to hear Europe, Africa, or the Middle East, but the odds of hearing Asia or the Pacific region will be poor. Listeners along the west coast have good possibilities for Asia, the Pacific, and Australian reception, but Europe and Africa will be very difficult. In both cases, listeners closer to the ocean itself will have better reception. BCB DXers in the midwest theoretically have a chance to hear all such areas, but their odds of hearing *any* are poor (although it has been done; some BCB DXers in the middle of North America have managed to hear all continents).

Some European and African stations operate all night, and you can try for them from about your local sunset until sunrise at the station (usually around 0530 to 0700 UTC). Other stations sign off around 0000 UTC and resume operations in the 0530 to 0630 UTC period. By contrast, DXing Pacific, Asian, and Australian stations is an early morning affair. The first stations from such areas do not begin to fade in until around 0800 UTC, and a possible propagation path to many areas does not exist prior to 1000 UTC.

Foreign BCB DX is primarily a fall and winter activity. The

lower ionospheric absorption, coupled with longer periods of darkness, means that many stations are realistically only possible then. QRN from thunderstorm activity is also reduced compared to spring and summer months. Late September to early November often has excellent conditions for European, African, and Middle East reception, while March and April are often the best months for the Pacific, Asia, and Australia.

Recent years have not been kind to the hobby of BCB DXing, particularly for those chasing domestic stations. The increased numbers of AM stations, and resulting QRM levels, have caused several BCB DXers to abandon the hobby altogether. Some of them have been among those recognized as the most skilled and experienced BCB DXers; apparently, they have literally run out of new countries and stations to hear with any reasonable amount of effort. Clubs specializing in BCB DX give some evidence of this trend, as material describing the "good old days" of BCB DXing appears in their pages with increasing regularity. However, there is a possibility that the days of "wide open" frequencies with few stations could be briefly repeated in the near future. The FCC has announced its intention to expand the BCB by an additional 100 kHz, to 1700 kHz, sometime within the 1990s. The expansion awaits formal agreements with neighboring countries, but already such firms as Motorola are testing antennas and transmitters for this new range. Current plans call for local and regional stations to fill the new range beginning at 1610 kHz. Even a modestly powered station on such frequencies could be heard coast to coast if QRM is low, as it will be in the beginning. In addition, there will be numerous tests at hours favorable for distant reception. Alert, active DXers during the first few months when this new range is opened will have the chance to make some "once in a lifetime" receptions which will become impossible as more stations crowd the new segment. It might be worthwhile to hone your BCB DXing skills and equipment in preparation for this opportunity!

FM and TV DXing

Many of those who gave up BCB DXing turned their attention to chasing DX on the FM broadcast band and TV channels. Success in FM and TV DXing often depends more upon being on the right frequency at the right time than it does upon receiving equipment or even skill.

Like BCB DXing, your location will have a big impact on what you can hear. Instead of east coast versus west coast, however, DXers in metropolitan areas with numerous FM and TV stations will be at a disadvantage, since local stations will block DX signals on the same frequencies. Those in suburban or rural areas will find the DX prospects much better. In these areas, a portable FM radio or television set can provide plenty of DX, even if a simple indoor antenna is used.

One propagation method for FM and TV signals is sporadic-E, which was discussed in chapter 5. While it can happen any time of the year, it's most common from late May until early August, with another period often occurring in late December. Sporadic-E is a very democratic mode of propagation; large outdoor antennas are often less effective than simple indoor "rabbit ear" antennas. Signal strengths via "E-skip" are usually strong, although frequently interrupted by abrupt, deep fading and distortion.

The first sign of a sporadic-E opening is usually "rolling" black bars across the picture received from a local TV station on channel 2, 3, or 4. These bars are caused by distant stations being propagated via sporadic-E; these stations are not strong enough to override the local stations but do cause the visible QRM. You may also hear some audio distortion as well. Sporadic-E is first noticed on channel 2 and "rises" to higher TV channels and the FM broadcast band as the ionization of the sporadic-E clouds increases. Sporadic-E propagation higher than television channel 6 is very rare. One of the fascinating aspects of sporadic-E is how the MUF will sometimes hit the middle of the FM broadcast band, such as 96 MHz. The result is that the FM band below 96 MHz is full of signals from hundreds or over a thousand miles away, while only local stations are heard above 96 MHz.

Reception during an E-skip opening can be wild and unpredictable. It's not unusual for a signal from a distant TV station to abruptly and completely fade away and be replaced by a signal from another distant TV station. A few minutes, another complete and total fade will bring the first station back. On FM, stations will surface above the jumble for just a few seconds and then be replaced by another station. Since sporadic-E clouds are in motion, the "direction" of an opening can change as the clouds move. (It's possible to track the motion of these clouds as stations on the same frequency but in different areas fade in and out; most sporadic-E clouds move in a westerly direction.)

Sporadic-E is most likely to occur from about 8 to 11 a.m. and from 4 to 8 p.m., your local time, although openings can happen any time. Some strong and intense openings begin as early as sunrise and can continue until well after midnight. E-skip is also more common the closer you are to the equator. Thus, listeners in the southern half of the United States will have an advantage over those in the northern half and Canada.

Sporadic-E is not the only propagation mode for distant TV and FM signals. Another is "tropo," which refers to propagation via *ducting* in the Earth's troposphere. The troposphere extends from the Earth's surface to an altitude of approximately six miles. This is the layer of the atmosphere where weather takes place; rain, wind, and storms are all products of the troposphere. Normally, the troposphere has no effect on radio signals of any frequency. But a very curious effect can be noticed on signals above 50 MHz. The temperature of the troposphere usually drops as altitude increases. But sometimes a layer of cool, dry air close to the Earth's surface is overridden by a layer of warmer, more moist air. This means that, at a certain point, the air temperature actually *increases* as the altitude increases. The situation is known as an *inversion* and is commonly found along weather fronts or large bodies of water. When inversions are present, signals above 50 MHz can get "trapped" at the point where the inversion is present and prevented from traveling out into space. Those signals then follow the curvature of the Earth, often for distances of several hundreds of miles.

Unlike sporadic-E, the effects of tropo are more pronounced as the frequency of the signal increases. The FM broadcast band is more often affected than TV channels 2 through 6, and channel 13 is more often affected than the FM band. In fact, there are some TV DXers who chase DX signals from hundreds of miles away on UHF channels thanks to tropo propagation.

Tropo lasts much longer than sporadic-E propagation, often for days at a time. Signals are not as strong as during sporadic-E reception, although they are more steady. Inversions typically form along slow moving fronts where cool, dry Pacific air meets warm, moist Gulf of Mexico air. Tropo is most common during the fall months, when warm, hazy days are followed by cool nights. Inversions also trap pollutants close to the Earth, so hazy, smoggy, and foggy weather is often a good indicator of good tropo conditions. Short range tropo openings of less than 300 miles can also occur around sunrise as the sun first warms the upper layer of the troposphere before the lower layers; such openings usually vanish within an hour after sunrise.

Successful tropo DXing generally requires more elaborate receiving equipment and antennas than E-skip. Some sort of rotatable outside antenna is a big help, as is a mast-mounted preamplifier. Good, selective TV sets and FM receivers are also a big help.

Tropo paths generally cannot form over mountainous areas. Listeners in Rocky Mountain states and provinces will find tropo rare, and tropo across the Rockies (as from California to eastern Colorado) is extremely rare. The Appalachian Mountains are not high enough to seriously disrupt a tropo path, so a tropo opening is possible in North America between most points east of the Rockies.

Amateur "Ham" Radio

SWLs often get confused with amateur radio operators or, as they are more commonly known, "hams." (Some SWLs have been known to pass themselves off as hams!) Hams are licensed to *transmit* on various frequencies from 160-meters (1800 to 2000 kHz) all the way up to the upper UHF ranges. In the United States, the necessary transmitting licenses are issued in different "classes" by the Federal Communications Commission to those who pass an examination for each class of license.

So what do hams do when they transmit? They talk to each other—using CW, AM, FM, SSB, RTTY, SSTV, TV, and even a computer-to-computer technology known as "packet radio"—at distances ranging from across town to the other side of the world. Some hams participate in on-the-air contests, trying to see who can score the most points by contacting as many different stations, countries, states, radio "zones," or even different call sign prefixes. Some hams swap QSL cards with other hams they contact, and try to earn awards for contacting all states, over 100 different countries, or even all counties in the United States. (There are over 3000 counties in the United States, and several hams have managed to contact and swap QSL cards with every one.)

There's a serious side to ham radio. In many major emergencies, such as the 1989 San Francisco earthquake, ham radio is the only two-way radio link which survives such disasters and is available to provide communications. Other hams provide auxiliary communications for special events such as parades and running marathons which can overload a city's normal communications networks. To keep in practice for such situations, some hams participate in emergency drills or networks which relay and deliver messages for free. A valuable

FIGURE 9-3

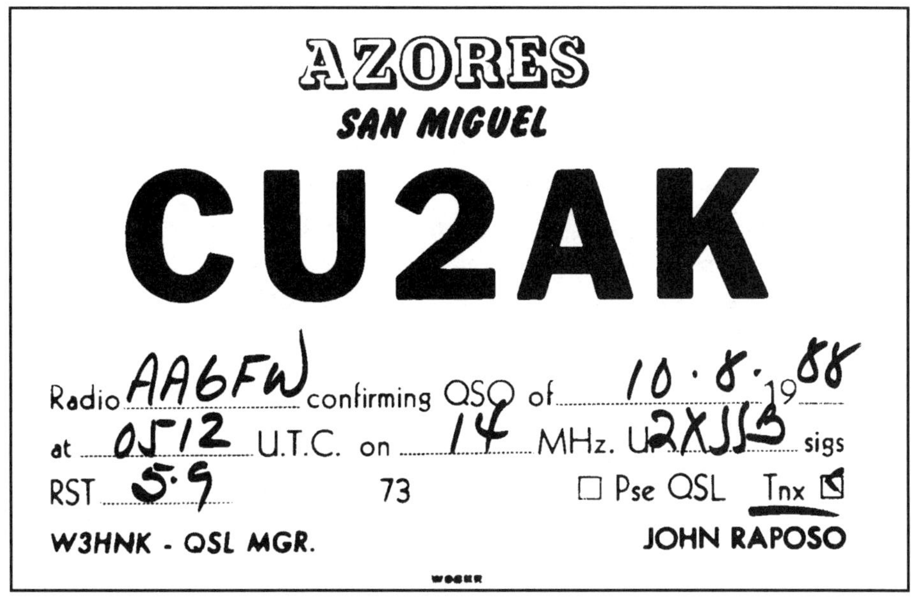

Ham radio DXers, like their SWL counterparts, are fascinated by QSL cards from faraway places!

service performed by hams in tornado-prone areas is "weather spotting." When conditions for violent weather are likely, hams travel to scattered observation sites and take along "walkie talkie" sets or use the ham stations installed in their cars to report any tornados or other dangerous weather (hail, heavy lightning, and the like) to the National Weather Service.

Ham radio operators have even constructed and orbited (with the assistance of various governments) their own series of satellites for amateur communications. The best-known of these satellites belong to the OSCAR (*o*rbiting *s*atellite *c*arrying *a*mateur *r*adio) series. The first OSCAR satellite was launched in 1961; the latest OSCAR satellites allow reliable, stable communications over wide areas of the Earth. In fact, the first satellite communications between the Soviet Union and the

United States involved ham radio stations K2GUN and UP2ON on December 22, 1965, using the OSCAR IV satellite. In fact, some American and Soviet space travelers have been licensed amateur radio operators and have operated while in orbit, delighting Earth-bound hams with the chance to talk directly to someone in outer space. (Actually, ham radio was represented from the beginning of manned space flight, since the first person to travel into space, Yuri Gagarin, was a ham. He didn't operate from space, however.)

But for many people the biggest attraction remains the ability to communicate with other people from all walks of life from all around the world. Hams are drawn from the entire human spectrum, meaning they're just as interesting or boring (or sane versus nuts) as any other segment of the general population. Tune across the voice segments of the ham bands in table 9-1

TABLE 9-1 Major Amateur Radio Bands and Their Uses

Band	Frequencies in kHz	Main Activities
160 meters	1800 to 2000	CW, RTTY, voice (usually LSB)
80 meters	3500 to 3750	CW and RTTY
	3750 to 4000	Voice (usually LSB) and SSTV
40 meters	7000 to 7150	CW and RTTY
	7150 to 7300	Voice (usually LSB) and SSTV
30 meters	10100 to 10150	CW and RTTY
20 meters	14000 to 14150	CW and RTTY
	14150 to 14350	Voice (usually USB) and SSTV
17 meters	18068 to 18110	CW and RTTY
	18110 to 18168	Voice (usually USB) and SSTV
15 meters	21000 to 21200	CW and RTTY
	21200 to 21450	Voice (usually USB) and SSTV
12 meters	24890 to 24930	CW and RTTY
	24930 to 24990	Voice (usually USB) and SSTV
10 meters	28000 to 28300	CW and RTTY
	28300 to 29700	Voice (usually USB and FM) and SSTV

and you'll soon see the truth of this. Some hams seem unable to converse on any topic other than the weather or the equipment they are using (and that suits many like-minded hams just fine), while others have truly interesting experiences to share. Where else can you meet and talk with a sound technician at the Universal Amphitheater in Los Angeles, a retired petroleum exploration geologist who has lived on every inhabited continent, a U.S. Navy submarine officer, a New York cable television producer, and a 12-year-old who wants to be a scientist—all in the same evening?

Moreover, amateur radio often melts away racial, cultural, class, religious, and national distinctions. Operating skill and acumen can't be bought, and status and privileges in the form of higher classes of operator licenses are open to anyone who can pass the requisite tests. Hams who try to converse with ("QSO") stations in foreign countries find themselves forced to become somewhat "internationalist" in their outlook and attuned to the political and cultural sensitivities of different nations.

The greatest growth area in amateur radio since the early 1970s has been in more localized communications, however. This is the area of *repeater* operations. A repeater is a relay station which listens for stations on one frequency (the "input" frequency), amplifies, and retransmits such stations on a different frequency (the "output" frequency). Repeaters mainly operate in the VHF and UHF ham bands, where communications are normally restricted to the horizon (or "line of sight"). Repeaters are placed in favorable locations, such as mountains or atop tall buildings, to increase the communications range of VHF and UHF stations, particularly those in cars or hand-held portable units. It's not unusual for two hams over 100 miles apart to communicate with each other using walkie-talkies through repeater stations.

To get a ham radio license from the FCC, you'll have to pass a written examination on radio theory and operating

FIGURE 9-4

CHIBA JAPAN ZONE 25

OP. TAKU NAKATA

QTH: 7-34-32 Makuhari-hongoh
Chiba-city Chiba 281 Japan

JR1IJV

Loc : QM05AQ

ALSO: WJ6L

TNX WW RTTY TEST

Confirming our QSO

TO RADIO	DATE	UTC / ~~JST~~	MHz	2WAY	RST
AA4FW	SEP. 23. 1989	2256	28	RTTY	599

RIG ~~TS-820~~ + 1KW Amp.
Ant 7ele Yagi, CQ, GP, DP, IV

PSE QSL via JARL or Direct
TNX NICE QSO CU AGN

Many hams use interface units with their personal computers to permit RTTY operation. Here's a QSL card for a contact with a Japanese station during an on-the-air RTTY contest sponsored by CQ magazine.

regulations. The difficulty and length of the written exam varies with the class of license you're applying for. The biggest stumbling block for most would-be hams is the current (as of the time this book was being written) requirement for passing a test on receiving messages in Morse code. Currently, international treaties require all applicants for an amateur radio license permitting operation below 30 MHz to pass a Morse code receiving test. International regulations permit individual governments to waive the Morse code requirements for amateur licenses permitting only operation above 30 MHz. Several nations—such as Great Britain, Australia, the USSR, and Japan—have already instituted "code-free" licenses themselves, and these have done a good job of attracting technically-oriented persons who have no interest in Morse code communications. At the time this book was being written, the

FCC was considering establishing a code-free class of ham radio license in the United States, and it appears likely that some form of one will be available by the time you read this.

The entire subject of Morse code requirements for licenses provokes a very emotional reaction among many hams. In the beginning of ham radio, Morse code was the only practical means of communication and a knowledge of it was a true necessity. As various radiotelephone modes grew in popularity, the use of Morse code declined. Today, it is used only by a minority of hams—compare the number of signals in the CW and voice segments of the bands and see for yourself. Despite this fact, many American hams have resisted the establishment of a code-free license. Various reasons are given by these people; the most common is the supposed value of Morse code in emergency situations. (However, in every major emergency during the past two decades, such as the 1989 San Francisco earthquake and Hurricane Hugo disasters, virtually all emergency communications have used either voice or RTTY/packet radio modes!) The real reason these people oppose code-free licenses is a "I had to do it, so you have to do it even if there's no need to" attitude coupled with an apparent desire to limit the number of people holding ham licenses. Fortunately, an alarming drop in the number of new amateur licensees, particularly among younger people, is gradually overcoming such arguments. Don't be surprised, however, when you run into those who insist that "real hams" must know Morse code!

Fortunately, the Morse code requirements are not as difficult as they might seem at first (or as tough as some hams like to boast!). The speed requirement for the lowest grades of ham licenses is only five words per minute (abbreviated "wpm"). This is a very slow speed, meaning that on the average one Morse character (a letter or number) is sent in slightly more than two seconds. Learning Morse code at this speed is about as difficult as memorizing the alphabet or learning to count from zero to nine. It's so simple, in fact, that many youngsters of seven and eight have mastered the code at that speed and have their licenses. (The biggest problem those kids had was not in learning Morse, but reading well enough for the written tests.) Knowledge of Morse code is tested in one of two ways: correctly copying one consecutive minute of code from a five minute long message or by correctly answering seven out of ten questions about the content of the message. There are several Morse code courses available on cassette tapes or as software packages for personal computers.

The written exams are all in multiple-choice format, and are in increasing difficulty by license class. While some study and effort are necessary to pass the written exams, most questions directly relate to amateur radio rules, safety, equipment, antennas, and things you'll have to know anyway to properly and safely operate a station. Numerous study guides are available to help you prepare for the written exams, and most people can pass the exams for the lower classes of licenses after only a few hours of study.

The FCC currently issues ham radio licenses free and they're good for ten years; they can be renewed without any additional examination. Actually, a ham radio license comes in two "parts" (even though you get only one form). The *operator* portion allows you to operate any ham radio station up to the limit of the license class, while the *station* portion is a call sign for use at the station location indicated on the license as well as portable and mobile operation away from that location. Ham radio licenses are often referred to as "tickets," and it's not unusual for hams to be better known to each other by their call signs than their names!

The call signs of amateur operators usually reflect the license class held by the operator. All call signs are issued from standard international allocations (see the appendix for a complete list) and consist of a one or two letter prefix, a single digit, and a suffix of one to three letters. If a higher class of license is obtained, hams can trade in their existing call letters for a new, shorter set reflecting their new license class.

At the time this book was written, the *Novice* class license was the first license class offered by the FCC. It requires passing a 5-wpm code test and a simple 30 question written test. It allows Morse code operation in small segments of the 80, 40, 15, and 10 meter bands. In addition, Novices can operate SSB in the 28300 to 28500 kHz segment of 10-meters, and are allowed FM voice privileges in the 222 and 1200 MHz UHF ham bands. Passing an additional 25 question written exam allows Novices to upgrade to the *Technician* class. The Technician license gives all Novice privileges plus all ham privileges above 30 MHz. Of particular interest is the 144 to 148 MHz ("two-meter") band, which is the most popular one for local communications, and the 50-54 MHz ("six-meter") band, which is near TV channel 2 and offers the same sporadic-E DX possibilities. If you apply for a Technician license first, you'll have to pass both the Novice and Technician written exams. This applies throughout the licensing structure; if you apply for a higher class license, you must pass (or have passed) all written exams for the lower license classes.

Next up the licensing ladder is the General class license. This requires a Morse code receiving test at a rate of 13 wpm and an additional 25-question written test. It gives voice and CW privileges in portions of the 80 to 15 meter bands and access to the entire 160 and 10 meter bands as well as all amateur bands above 30 MHz.

The *Advanced* class license gives almost all voice privileges in the 80 to 15 meter bands. For a holder of a General class license, obtaining an Advanced class ticket means passing an additional 50-question written test on advanced radio theory and communications methods. At the top of the licensing ladder is the *Extra* class license, which requires passing a 20-wpm code test and a 40-question written exam on specialized communications topics.

While this licensing structure was in effect at the time this book was written, the FCC proposal to add a code-free license to the structure might mean other classes could be dropped or new requirements established for the rest. The latest details can be obtained from the FCC or from the American Radio Relay League (ARRL). The ARRL is the national association of ham radio operators, and is located at 225 Main Street, Newington, CT, 06111. The ARRL is a nonprofit organization which sponsors a variety of training programs, operating activities, and awards for hams and also publishes several books on ham radio. A note to ARRL will bring you membership and publications information.

The ARRL also publishes a monthly magazine called *QST*. Unfortunately, this magazine is almost impossible to find on newsstands. Easier to find is CQ, a commercial publication with an emphasis on DXing, contests, and other ham radio operating activities.

Tuning across the ham bands listed in table 9-1 will give you a good idea of the types of activities hams engage in. If you get interested in becoming a ham yourself, a good step would be to contact a local ham radio club. Many have training courses to prepare for the various exams; every one will have members willing to help beginners and to demonstrate ham radio. The names and addresses of clubs in your area can be obtained by dropping a note to ARRL and enclosing a self-addressed stamped envelope with your request.

If you do wind up getting your own ham radio license, be sure to listen for a station identifying itself as AA6FW. That's me, and I'd love to know how things are going with you!

CHAPTER TEN

Unusual, Illegal, and Mysterious Radio Stations

ALL THE DIFFERENT TYPES of radio stations that we've looked at so far have been "on the level." That is, all are operated legally (if sometimes covertly, as with military communications) under the authority of a governmental entity. But scattered throughout the shortwave bands are stations that are beyond the normal controls of governments and licensing authorities. Some are "pirate" radio stations, often operated by young people, that broadcast without bothering to get a license. Others are used by smugglers and guerilla forces to coordinate their activities. You can also hear women's voices reading groups of numbers in various languages; these are believed to be coded instructions to espionage agents.

There are also broadcasters operating from unknown locations who try to incite revolution in various countries or areas. And there are other types of strange signals which defy easy categorization. All of these can be heard—and usually with ease—on your shortwave radio.

The key to hearing much of the activity in this section is patience. These stations seldom operate on regular schedules and frequencies; instead, you'll have to carefully search for them. This means hearing these stations often depends more upon being in the right place at the right time than it does upon your equipment or even DXing skill. The latest information is an important tool. Membership in a SWL club or a subscription to a magazine like *Popular Communications* is essential to keep current on the many rapid changes that take place in such activity.

Clandestine Broadcasting

Clandestine broadcasting stations sound much like any "normal" shortwave broadcasting stations. However, clandestine stations differ because they are *deceptive*. Other international broadcasters may slant the news or even outright lie about events, but clandestines take deceptiveness to another level. For example, clandestine broadcasters often keep their true location or sponsoring organizations secret. A clandestine station might pretend to be broadcasting from within its target country when in truth it's operating from a country opposed to the target nation. Clandestines are overtly political creatures; they exist to bring about some political change or action in their target countries. When political conditions in the target or sponsoring country change, the clandestine goes off the air or a new clandestine station may go on the air. For example, southeast Asia was a hotbed of clandestine activity during the Vietnam War, but those stations largely vanished after the end of the conflict. More recently, Central America was the site of clandestines sponsored by the Sandinista government in Nicaragua and other stations opposing them were located in neighboring countries. With the electoral defeat of the Sandinistas in 1990, many of these stations left the air.

Clandestine broadcasting began in a big way at the start of World War II. Perhaps the first was Radio España Independiente, the voice of the exiled Spanish Communist Party. This station first took to the air in 1941 from transmitters in the Soviet Union, and continued to broadcast from various locations in the Soviet bloc until it closed in 1977 following the death of Francisco Franco. Radio España Independiente was in many ways the prototype for all future clandestine broadcasters. During its years of operation, it called itself "the station of

Pyrenees" (the mountains between France and Spain) and tried to convince listeners it was broadcasting "on the run" within Spain itself. After the political changes in Spain, the station was no longer needed and it vanished.

Perhaps the most famous clandestine broadcaster of all time was the now-legendary Radio Swan/Americas. In May, 1960, SWLs were startled to hear a new station calling itself Radio Swan on 1160 and 6000 kHz. All programming was in Spanish, with a strong anti-Castro slant, and the station claimed to be a commercial venture located on Swan Island in the Gulf of Mexico.

Some very peculiar aspects of Radio Swan's operation quickly became clear. Swan Island was claimed by both the United States and Honduras, and the FCC issued amateur radio licenses for operation there. However, the FCC claimed to know nothing whatsoever about Radio Swan. (This was despite clear evidence from direction-finding on Radio Swan's 1160 kHz signal that definitively showed the signal was coming from Swan Island.) The station was owned by a "Gibraltar Steamship Corporation," which, as it turned out, owned no steamships. The president of Gibraltar Steamship, Thomas D. Cabot, was the former head of the U.S. State Department's Office of International Security Affairs—a curious background for the head of a supposed maritime shipping company. The few commercials aired were clearly inadequate to support a commercial venture on the scale of Radio Swan. Fidel Castro himself was convinced the station was a covert Central Intelligence Agency operation; in a September, 1960 speech to the United Nations he spent four and half hours denouncing Radio Swan as an attempt to undermine his regime.

Any pretense that Radio Swan was merely a commercial station went out the window with the abortive Bay of Pigs invasion in May, 1961. Radio Swan transmitted coded messages to the invading forces and guerillas inside Cuba, suspending normal programs during the invasion and serving as a communications link for invaders. A few months after the failed invasion, the station changed its name to Radio Americas and continued operating until the spring of 1968. As Radio Americas, its programs consisted of many anti-Castro talks and newscasts along with pop music in English and Spanish.

Why Radio Americas was closed remains a secret, but apparently the CIA decided the station was not effective in its mission of undermining Castro. Just before Radio Americas was closed, well-known radio writer Tom Kneitel and the editor of the now-defunct *Electronics Illustrated* magazine were allowed to visit Swan Island under the guise of conducting ham radio operations. They found Radio Americas in full operation, and wrote a feature story about their visit for *Electronics Illustrated.* It hit print around the same time Radio Americas left the air for good. In retrospect, it seems clear the decision to close the station had been made some months before and that for some reason the CIA wanted Swan Island firmly established as the station's location. (Some SWLs believed this was because not all transmissions had always come from Swan Island, with Mexico's Yucatán Peninsula considered a likely spot, and the CIA wanted to protect any foreign nations in their dealings with Cuba.)

The patterns of Radio España Independiente and Radio Swan/Americas continue to be repeated in more contemporary clandestine efforts. One that was operating when this book was being written was Radio Truth, an anti-Zimbabwean clandestine sponsored by the government of South Africa. Radio Truth operates on 5014 kHz, and its English programs often put good signals into North America around 0430 UTC. Radio Truth does not announce a transmitter location, nor does it mention its links with South Africa. Instead, Radio Truth tries to pass itself off as just another domestic station broadcasting from "somewhere."

A station which seemed to borrow a page from the Radio Americas book was Radio Impacto, a supposedly commercial station located in San Jose, Costa Rica. Unlike Radio Swan/Americas, there was never any doubt where this station was—SWLs visiting Costa Rica were easily able to locate its studios and transmitter site—but there were many questions about its true sponsors and purposes. The building itself was surrounded by armed guards. The programming had a strong anti-Communist and anti-Sandinista slant, with strong support of U.S. policies in its newscasts and commentaries. Commercials were few and far between, certainly not enough to support an operation of its scope. Yet the funding was clearly adequate to support a high-quality broadcast operation that put strong signals into North America on 5030 and 5044 kHz during the evening. A couple of months after the electoral defeat of the Sandinistas, Radio Impacto abruptly left the air without explanation or warning. While the truth about Radio Impacto may never be known, it's likely that stations using its *modus operandi* will appear in the future according to world political conditions.

A major operation that was still continuing as this book was being written was Radio Caiman (Spanish for "alligator"). It first appeared in 1985, and was dubbed "Radio Nat King Cole" among SWLs because of the large number of records played that were by or in the style of that singer. It wasn't until several months later that it finally began identifying as Radio Caiman. The station added more contemporary music, special programs for teenagers and young adults (such as "La Hora de Juventud"), and slips in anti-Castro messages and quips in a very smooth, laid-back style. Radio Caiman has a professional sound, stable frequencies, and no announced sponsor or backing group. At the time this was written, Radio Caiman was heard well on 9965 kHz in the mornings and evenings.

A second anti-Castro clandestine active when this book was written was La Voz de Cuba Independiente y Democrática (CID). This is sponsored by the Cuba Independiente y Democrática organization under the leadership of Huber

Matos, Jr. In the past, this station has been relayed over some legitimately licensed broadcasters (such as Radio Clarin in the Dominican Republic) but also used several undisclosed transmitter sites. When this book was written, it could be heard around 9941 and 7340 kHz (both frequencies variable) during evenings with good signals strength.

Southeast Asia is another hotbed of clandestine activity. In Sri Lanka, the Tamil National Broadcasting Station and the Voice of Freedom contend with each other for the minds of that nation's Tamil population. The former is believed to be operated by the Indian government, which sent troops to Sri Lanka in 1989, while the latter is believed to be operated by the Sri Lanka government itself.

After the events in China during 1989, it was inevitable that some sort of clandestine activity directed toward Beijing would begin. The best-heard efforts so far have been the Voice of June Fourth, sponsored by Chinese students and exiles abroad. This is broadcast over transmitter facilities provided by the government of Taiwan, although initially attempts were made to keep this involvement a secret. As you might expect, this station is heavily jammed by the Beijing government.

Another Asian clandestine is the Voice of the Khmer, operated by a Cambodian resistance group organized by deposed Cambodian ruler Prince Sihanouk. Although programs are entirely in Cambodian, many SWLs in North America have heard it on 6325 kHz around 1100 to 1200 UTC.

The two Koreas have been aiming clandestines at each other for some time. One interesting twist is how some of these stations have pretended to operate from the capital of the target country! Listeners in North America have reported good reception of the Voice of National Salvation on 3480 and 4450 kHz around 1130 UTC in Korean; this station is operated by the South Korean government.

The Middle East and northern Africa have also seen much clandestine activity in recent years, with Iran, Ethiopia, and

Libya being some of the primary targets. These stations are run by exile groups and broadcast from countries currently at odds with these nations; Egypt has been one of the major providers of facilities for such stations, particularly for those directed against Libya.

The rapidly changing clandestine radio scene can best be followed through the "Clandestine Communique" column each month in *Popular Communications* magazine or by membership in a SWL club that covers clandestine topics.

Pirate Radio

Like clandestines, pirate stations are secretive broadcasters who often employ a great deal of deception in their operations. Unlike clandestines, however, pirate stations generally have no political ax to grind. They are on the air for the same purpose as your local AM and FM stations—to provide entertainment and amusement. But, unlike your local broadcasters, pirate stations don't broadcast to make money. Nor, for that matter, have they bothered to get a broadcasting license!

Pirate radio got started in Europe, where broadcasting was (and largely still is) a government monopoly. Radio broadcasting was restricted to a very limited number of government-run networks (such as four in all of Great Britain) and the programming tended to be what government bureaucrats felt people needed to hear instead of what they wanted to hear. The rock and roll music explosion of the 1950s and increased immigration from Third World countries created a demand for alternative types of programming that the existing government networks did not respond to.

The first European pirate stations were all commercial ventures operating ships in international waters. The first "sea pirate" was Radio Mercur, which began operations in 1958 with programs for Denmark. 1964 saw numerous pirate broadcasters begin operation off the British coast. Stations such as Radio

Caroline, Radio London, Radio Scotland, Radio 390, Radio City, and Britain Radio all used a format similar to American "Top Forty" radio. For the next three years, these stations provided a viable alternative to the BBC's pop music network. In 1967, however, the British Parliament passed "The Marine Broadcasting Offences Act," which made it illegal to advertise on, work for, supply, or otherwise assist an offshore broadcaster. Most offshore broadcasters closed down, although a few (such as Radio Caroline) responded by moving their offices and supply points to the Netherlands.

In an attempt to defuse the "demand" for pirate broadcasters, Britain permitted a limited degree of private commercial broadcasting. Other European nations established new government-run networks with limited advertising. But these steps did not satisfy some of the listening public. Partly this was because the number of private and commercial broadcasters remained low, and the variety of programming suffered as the stations tried to be all things to all people. (An example was London, which for years had only two private broadcasting stations for the entire city.) Those interested in music such as jazz, soul, and country still found a limited amount of broadcasting for them. Moreover, there were some onerous restrictions placed upon the "independent" broadcasters, such as a requirement that no more than 50% of the programming could be music. As a result, a new wave of pirates sprang up, mainly in Britain and the Netherlands, during the 1970s. Unlike the seagoing pirates, these were land-based, nonprofit, and operated almost as a public service by young people. Some, such as Dred Broadcasting, London Greek Radio, and Voice of the Immigrants, broadcast to sections of the population neglected by the government networks. A few of these pirates were operated on shortwave and were heard in the United States, and were the subject of considerable interest among SWLs.

There was very little land-based pirate activity in North America prior to the early 1970s. Most stations which did operate prior to that period were either one-shot affairs (such as Radio Free Harlem, a joke by a ham radio operator in the 1960s that was—much to the operator's chagrin—taken very seriously by the authorities) or "kids playing radio." The latter stations were quickly closed down by the FCC or Canada's Department of Communications. The precise factors which resulted in the explosive growth of pirate radio in the 1970s are difficult to determine, but two major ones seem to have been the availability of old AM amateur radio transmitters at bargain prices and a perception that the FCC was unable to enforce its regulations, as demonstrated by the chaos that swept over the CB channels in the mid-1970s. Potential pirates began to perceive, not entirely without good reason, they could operate with relative impunity as long as they avoided interference to other stations.

Pirates since the 1970s have tended to congregate in two unofficial "pirate broadcasting bands." One is the range just above the standard broadcasting band from 1600 to 1650 kHz; this range is particularly favored by pirate broadcasters in the New York City area. A second band is from 7400 to 7500 kHz and is used by pirates throughout the United States and Canada. Several stations operate around 7415 kHz. Most transmissions are in AM, and the usual operating times are during the evenings and nights on weekends and major holidays. (Halloween has become a favorite operating day, with several pirates usually active then.)

One of the most fascinating aspects of current North American pirate activity is the many links these stations have to the SWL hobby; many are operated by SWLs. (Many SWLs who have reported reception of pirates in various SWL club bulletins, but not directly to the station, have been surprised to find QSLs arriving from those stations a few days after the club bulletins were mailed.) The prototype for such stations was the "Voice of the Voyager," which exploded on the scene in early 1978. This station was located near Minneapolis, Minnesota, and was operated by several teenage SWLs. It chose 5850 kHz

FIGURE 10-1

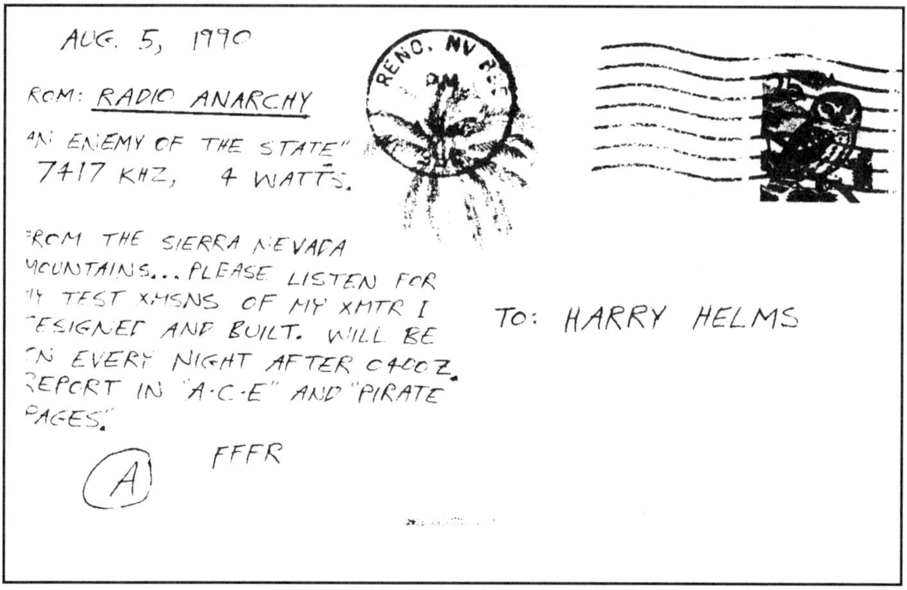

AUG. 5, 1990

ROM: RADIO ANARCHY

AN ENEMY OF THE STATE"
7417 KHZ, 4 WATTS.

RCM THE SIERRA NEVADA
YCUNTAINS... PLEASE LISTEN FOR
Y TEST XMSNS OF MY XMTR I
ESIGNED AND BUILT. WILL BE
N EVERY NIGHT AFTER 0400Z.
EPORT IN "A-C-E" AND "PIRATE
PAGES."

(A) FFFR

TO: HARRY HELMS

An example of how pirate broadcasting is linked to the SWL hobby—advance notice of an upcoming pirate broadcast. Several other SWLs along the west coast received similar postcards.

as its frequency, and began late Saturday night broadcasts which quickly attracted a cult following among SWLs in North America. In addition to the inevitable rock music, the operators took phone calls from SWLs, ran satirical items about the FCC and SWLing, and even gave out the latest SWLing and DX news. This 100 W pipsqueak likely had a larger, more loyal audience in North America during 1978 than most international broadcasters!

Unfortunately for its operators, the Voice of the Voyager was raided by the St. Paul, Minnesota, office of the FCC in August, 1978, and the station was taken off the air. The FCC only issued a warning to the operators, however, and the tone of the "bust" was surprising cordial; the FCC agents requested—and received—Voice of the Voyager QSL cards!

The station did not remain silent for long, however. It

returned to the air and continued to operate sporadically well into the early 1980s, leaving the air for lengthy periods due to equipment problems and a shortage of participants. The FCC raided the station a second time, and on this occasion there were no warnings or requests for QSLs—instead, fines of several hundreds of dollars were handed out to each of the major participants.

While the Voice of the Voyager itself was eventually silenced, the seed had been planted, and pirate broadcasting was off and running in North America. The FCC and Canada's Communications Canada have made attempts to stem the tide by periodic crackdowns involving hefty fines and confiscation of station equipment. All of these efforts have been heavily publicized. In spite of this, pirates still continue to operate and will likely continue to do so in the future, since the FCC cannot monitor every single frequency and transmitting equipment can be freely purchased and possessed even by those lacking an appropriate station or operator license.

The business of pirate broadcasting has become controversial within the SWL hobby and clubs, with listeners divided on whether acknowledging these stations even exist encourages them. A few SWL clubs have responded by banning any information about pirates beyond the basic details (date, time, frequency, etc.) concerning receptions of them and by ruling that QSLs from such stations do not count toward club awards and certificates. (This policy is not followed in regard to European pirates, which are every bit as illegal in the countries from which they operate.) Other clubs are more tolerant, and one specialty club, the Association of Clandestine Radio Enthusiasts, has been established.

While most pirate broadcasters in North America seem to be in the Voice of the Voyager mold, a few are apparently more serious operations. One is the Voice of Tomorrow, which made its first broadcast in 1983. This station has superb, professional production values, and its signal is usually strong throughout

FIGURE 10-2

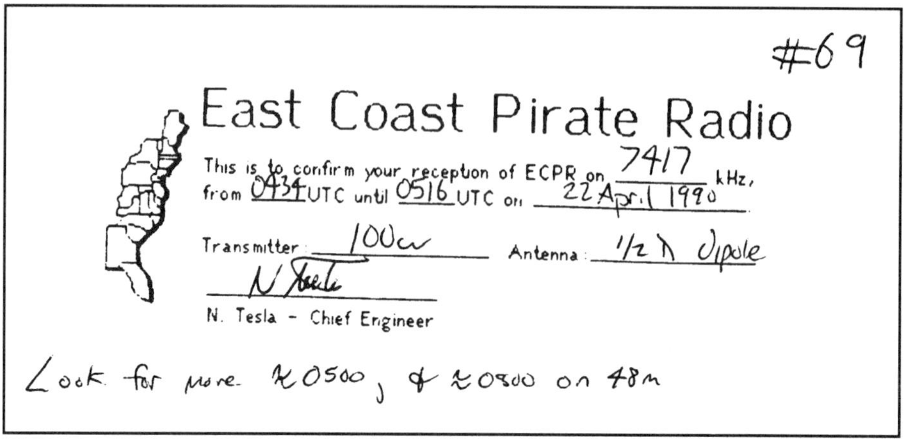

#69

East Coast Pirate Radio

This is to confirm your reception of ECPR on 7417 kHz, from 0434 UTC until 0516 UTC on 22 April 1990

Transmitter: 100w Antenna: 1/2 λ Dipole

N. Tesla - Chief Engineer

Look for more ≈0500, & ≈0800 on 48m

Pirate stations will also send out QSL for correct receptions reports. This is one from East Coast Pirate Radio.

North America. The most remarkable thing about this station is its programming; it is a neo-Nazi broadcaster whose programs are consistently anti-Semitic and anti-black, adopting a consistent white supremacist view. There are often lengthy political speeches and interviews, all with an explicitly fascist theme. Despite the explosive subject matter, the presentation style is calm and low-key. The station signs on with an interval signal consisting of a howling wolf over a steady drumbeat, and most broadcasts last from an hour to 90 minutes. While operation is on an irregular basis, it is usually heard around 6240 or 7410 kHz in the 0030 to 0300 UTC time frame. Many SWLs are curious as to why this station has been able to operate with relative impunity by the FCC for so many years, especially considering its political content.

Another pirate station whose content is less threatening but still interesting is the Voice of Bob. This station takes its programming from the Church of the Subgenius, a Dallas-based satirical group whose targets include almost all organized religions but with a special disdain reserved for fundamentalist evangelists. The Church of the Subgenius is based around the veneration of J. R. "Bob" Dobbs, a cartoon character who strongly resembles the late Hugh Beaumont, who played Ward Cleaver on the "Leave it to Beaver" television series. The Church of the Subgenius produces radio programming which is a favorite of college stations across the country, and this programming has sometimes been relayed over a pirate station calling itself the Voice of Bob. This station has operated since 1984, but only on an irregular schedule.

Whatever your feelings, pirate broadcasting seems to be here to stay in North America, and many of these stations are challenging DX. Indeed, trying to hear and extract a QSL from an illegal, low-powered station can be a major challenge. Membership in a SWL club that covers pirate activity is a must; in addition, there is no substitute for careful tuning and listening for pirates, since most operate erratically.

Numbers Stations

For years, SWLs have stumbled across transmissions that consisted of little more than a voice, usually that of a woman, reading numbers in groups of four or five digits in such languages as Spanish, English, and German. These have been the subject of much speculation in the hobby, but today the evidence is overwhelming that such transmissions are actually coded messages to various espionage agents operating in different countries. These have become known as *numbers stations* by SWLs.

It doesn't take a great deal of effort on your part to hear these stations. If you tune outside the broadcasting and amateur bands from approximately 0000 to 0800 UTC in North America, you'll run across several such stations, usually in Spanish. While you can hear numerous stations using four and five digits, the five-digit stations are more plentiful. On some of the five-digit stations, you'll sometimes hear a distinct pause between the third and fourth digits; these have become known as "3/2-digit" stations. In addition to Spanish numbers, you'll also hear groups in English, German, and "tone-keyed" Morse code. These latter stations really stand out, since they are AM stations over which the Morse code is sent using audio tones. Such "Morse" stations can be heard without using your receiver's BFO.

Numbers stations almost always use a female voice to read the numbers, although on rare occasions a man's voice might be used. (It's not known if there is any special significance attached to the use of a man's voice.) Most transmissions are in AM, although SSB has been used, particularly on German language stations. (On many of the AM stations, only one sideband is transmitted with the carrier; the result is full compatibility with AM receivers and improved transmitting efficiency.) You'll notice that the various digits making up a message have the same inflection and sound—a "seven" in the first group of a message will be identical to a "seven" in the last group. This is because all these messages are composed from a recorded set of digits from 0 to 9 along with a few words used to open and close each transmission. The effect is very similar to listening to a telephone recording for incorrect and discontinued numbers, and it's likely that similar equipment is used to prepare numbers stations messages. In rare cases, a few apparently "live" numbers transmissions have been heard (once, an announcer paused to cough) but these are very much the exception.

Some SWLs have attempted to decode these various messages, but this is an exercise in futility. The encoding system used for most messages is apparently the *one time pad* method. In this system, the intended recipient of the message has a copy of a pad with columns of numbers, in groups of four or five digits, printed on it. At the beginning of the message, the sender transmits the *key* to using the one-time pad. The key indicates which page, column, and line the recipient should turn to. The key also tells the recipient whether the number groups in the message should be added to or subtracted from the number groups printed on the pad. After the message is transmitted, the recipient takes the copy of the message and performs the required addition to or subtraction from the number groups printed on the pad. Each number group usually stands for a segment of the message, such as "at noon," rather than individual digits representing letters of the alphabet. To confuse unwelcome listeners, meaningless number groups can be inserted into messages at random. This system sounds simple, and it is compared to most methods. However, its security is close to 100% unless copies of the appropriate one-time pads are obtained by outsiders. The principal problem with the one-time pad method is the time it can take to decode a message. Encoding is simple thanks to computer-based systems.

Another encryption method believed to be in use with the 3/2-digit messages is the *dictionary key* system. In this, both the sender and recipient agree to use a book available to both (such as a dictionary, although it could be any book) as the key to decoding a message. The first three digits represent a page number, and the last two digits indicate the position of a word on that page, such as counting from the upper left of a page. By going to the pages indicated in a book and locating the appropriate word, the entire message may be reconstructed. This method has the advantage of not requiring the recipient to have one-time pads or similar incriminating evidence in their possession; it also means contact could be maintained by radio with long-term "sleeper" agents without personal contact or travel by the agents. The downside is that such a code can be easily compromised if the "key" book is discovered; several years of messages could be broken if that were to happen.

The two "brands" of numbers transmissions (four-digit and five-digit) sign on in different ways for Spanish messages. For five-digit groups, the opening is usually the word "atención" followed by a three-digit group and a two-digit group, as in "atención 545 49." The meaning of the three-digit group is unclear; it could be the intended recipient of the message or the key (the page in the one-time pad book or a location on a page, for example) used to decode the message. The two-digit group is always the number of groups composing the message, so in this example there would be 49 five-digit groups making up the message. This sign-on "announcement" is repeated for several minutes before the actual message begins. After the message is transmitted, the words "fínal, fínal" end the message.

Four-digit Spanish stations normally sign on with a three-digit group followed by a count (in Spanish) of "1, 2, 3, 4, 5, 6, 7, 8, 9, 0"; this pattern is repeated several times. The purpose of the three-digit group is uncertain, but it's likely the intended recipient or the key to the message. After this opening pattern has been repeated, the next item is the word "grupo" followed by the number of groups making up the message. This is usually repeated a few times. The four-digit blocks making up the message follow, and the entire message is normally sent twice. Four-digit stations leave the air as soon as the message is sent a second time without any sort of announcement.

Numbers stations in other languages follow similar formats, with five-digit stations using "attention" or "achtung" in place of "atención," while four-digit transmissions open with a three-digit group followed by a two-digit group count. Sound effects, such as tones, beeps, and even music, may also be used.

At first, it might seem that numbers stations would always operate in an erratic and unpredictable manner. That isn't the case. Certain frequencies, such as 5810 and 6840 kHz, have been used for years by numbers stations. Transmissions may be repeated within an hour of the first transmission. A few messages are repeated, digit for digit, days and even weeks after they're first transmitted. Moreover, many transmissions, particularly of the five-digit variety, tend to be found in certain frequency ranges depending upon the three-digit key or identifier. Table 10-1 shows some typical frequencies used by numbers stations, although random tuning during the evening and night hours will produce many other frequencies.

There are some SWLs who doubt that numbers stations are

used for espionage communications, and instead believe they could be used for transmission of such things as business data or even lottery numbers. (Curiously, a couple of employees at a FCC monitoring station put forth such theories in an interview conducted in an issue of the Association of Clandestine Radio Enthusiasts bulletin!) But the evidence for these being espionage communications is large and from a variety of sources. For example, in the book *The Spy Who Got Away*, author David Wise tells how CIA defector Edward Lee Howard was taught, as part of his CIA training, to receive messages transmitted as five-digit groups. John Barron describes in *KGB Today* how several different Soviet agents have received messages from the Moscow KGB headquarters as five-digit numbers groups in

TABLE 10-1 A Sampling of Numbers Stations Activity

kHz	Description and Times
4030	Five-digit Bulgarian and Serbo-Croation groups 0500, 0630
4882.5	Five-digit Russian groups 0540
5750	Five-digit German groups 2200
6515	Five-digit Spanish numbers 0230
6520.6	Five-letter CW groups 0200
6840	3/2-digit English groups 2340
	Four-digit Spanish groups 0230
6890	Five-digit Spanish groups 0700
6970	Five-digit English groups 0500
7532	3/2-digit German groups 2230
7763	Four-digit English groups 0130
8056	Five-digit Spanish groups 0300
8089	3/2-digit English groups 0215
8165	Five-digit Spanish groups 0640
9273	3/2-digit English groups 0400
11490	Four-digit Spanish numbers 1830
12242	3/2-digit English groups 0315
13649	Five-digit English groups 0430

Morse code. The paperback edition of *The Puzzle Palace* by James Bamford carried a new chapter on the trial of English spy Geoffrey Prime, who was arrested with one-time pads in his possession and received instructions as five-digit groups in English from a transmitter in East Germany. And in *Widows*, by Joseph and Susan Trento and William Corson, there is a wealth of detail about how various spies have been sent messages as number groups by shortwave radio.

The question of where the various numbers stations are located has long been controversial. For years, it was assumed that the five-digit stations were in Soviet Bloc nations such as Cuba, East Germany, and the Soviet Union while four-digit stations were operated by the United States and its allies. In the past, the FCC sometimes replied to queries about the five-digit Spanish numbers by stating they were in Cuba. In other cases, the FCC muddied the waters by claiming to know nothing whatsoever about numbers stations.

A few listeners claimed to have traced numbers stations to their sources. Back in 1978, I wrote an article on numbers stations for a national electronics magazine and received a letter from a reader in southern Florida. This reader claimed he had used a portable SW radio to locate a numbers station transmitter site in southern Florida. The letter had plenty of detail and the reader showed enough technical knowledge to make his story plausible. I wrote back, asking for more information and confirming details, but unfortunately never got any response. Similar stories circulated about other SWLs who supposedly found the transmitter sites for various numbers stations. Some were credible, while others (some SWL on a hike in the woods stumbles across a heavily-fortified radio station and is frightened away by rifle-toting guards) were much less so.

The first definite location for a numbers station was found in 1984, when a mathematics professor at a New England college visited one likely site armed with a portable SW receiver. The site was south of Washington, DC, near Warrenton, Virginia, on the Vint Hill Farm Military Reservation. This was one of the first sites used by the National Security Agency for intercepting radio signals; the site today (per a sign outside the facility) is part of the National Communications System of the U.S. government.

A major breakthrough in locating numbers stations was made in late 1989 by a former intelligence officer who writes about numbers stations under the pseudonym of "Havana Moon." He was able to secure the cooperation of an agency of the United States government to perform some direction-finding on various numbers stations. This agency was able to locate transmitter sites at Jinotega, Nicaragua for a five-digit Spanish station on 6577 kHz, at Guineo, Cuba for a five-digit Spanish and Morse code station on 3927 kHz, and at Havana, Cuba, for a five-digit Spanish station on 3690 kHz.

This same agency was asked to help locate several four-digit transmitters. Each time they were asked, however, they were unable to help for a variety of reasons. Perhaps one involved the discovery of a second four-digit transmitter site in the Miami area (near the zoo) in 1989. Signal strengths and fading patterns strongly indicate that additional numbers station transmitters may be located in the greater New York area, southern California, and possibly southern Ohio.

It should be clear by now why following numbers stations has such an attraction for many SWLs, including me! How often do you get to live out a James Bond-type fantasy and match wits with the KGB and CIA? If you have a portable SW radio, some patience, and live in areas such as greater Washington, DC or southern Florida, you might be able to make a major contribution to clearing up some of the remaining mysteries about these stations!

Phonetic Alphabet Stations

Phonetic alphabet stations are similar in many respects to numbers stations. They use a female voice to repeat a phrase composed of letters from the international phonetic alphabet, such as "kilo bravo alpha two." These stations can be repeated for hours at a time, but may be interrupted for messages consisting of groups of letters from the phonetic alphabet. These stations first began to be widely heard in the late 1970s.

An article by Greg Mitchell in the April, 1984 issue of *Popular Communications* magazine claimed these transmissions are produced by the Mossad, the Israeli intelligence service. Mitchell presented compelling evidence (based on a visit to Israel) that some of these stations do indeed transmit from Israel, and the Mossad theory is currently the most widely accepted explanation as to their origin. However, it's clear not all of these transmissions can originate from Israel. Some are heard at times and on frequencies unlikely to support propagation from Israel; some monitors in the metro New York City area have reported powerful, local-like reception indicating the transmitter was nearby. It's possible some transmission could originate from Israeli embassies and consulates around the world or from relay transmitters of friendly nations. Table 10-2 shows some of the more active frequencies for these stations.

Smugglers and Guerilla Groups

There's an enormous and growing amount of unusual, unknown, and shadowy activity taking place throughout the SW spectrum. Part of this is due to the technology of SW reception and transmission. High performance SW radios, smaller than a typical hardcover book, with direct frequency readout and SSB capabilities, are now available for about $200.00. The advances made in transmitting technology have been even more dramatic. Many amateur radio transceivers today are capable of transmitting *anywhere* from 150 kHz to 30 MHz with extremely simple modifications. (In the case of one very popular unit, this involves merely moving an internal switch to the opposite position; in other instances, the necessary modifications mean clipping or unplugging a couple of wires.) This has resulted in a

TABLE 10-2 Some Unusual Radio Signals

kHz	Description
4705	Pulses, one per second, 0500
4881	"Uniform Lima X-ray Two" repeated 2330
6676.6	Fishing vessels using USB 0615
6669.7	Spanish language communications using USB 0230; believed to guerilla forces in Central America
6735	"X" repeated in Morse code 0140
7394.5	"V" repeated in Morse code 0005
7445	"Kilo Papa Alpha Two" 0015
7692	Three dots, one dash rasper 0130
10125	"Charlie India Oscar Two" repeated 0245
10187	Eight dots, one dash rasper 0430
12950	"Victor Lima Bravo Two" repeated 0630

dramatic increase in the use of SW radio by drug smugglers, guerilla groups, pirate broadcasters, and similar people not likely to go through the formalities of a license, call signs, or established operating procedures.

Smugglers are some of the prime users of this new transceiving technology. Smuggling transmissions are likely to be heard anywhere throughout the SW bands, but the 6200 to 7000 kHz and 7400 to 8000 kHz ranges are particularly active. Most smuggling transmissions heard in North America take place during the hours of darkness, and involve narcotics brought in along the Atlantic and Gulf of Mexico coasts. You'll seldom heard an explicit reference to drugs (although I once heard a station talking about a load of "brownies" it had); instead, you'll hear remarks about meeting points, arrival times, dropping off items, and similar topics. Smuggling transmissions can be identified by their improper radio procedure (such as their lack of call signs and use of whistling), cryptic nature, and operation in incorrect frequency segments. You'll hear both English and Spanish used.

Not all modified transceivers are used by smugglers, however.

Many persons are using them in a quasi-amateur radio fashion to stay in touch with friends across the country and world. The 6700 to 7000 kHz range seems to have many of these stations. Some of these operators have formed groups which issue call signs to identify the various stations, and they can sound much like an ordinary ham radio contact when you first tune in. Commercial fishing vessels have also started using converted ham transceivers; these are identified by their constant references to the day's catch, drinking, and sex.

Guerilla forces in Central and South America are other major users of converted gear. Unfortunately, you'll have trouble understanding these communications unless you're fluent in Spanish but you won't have any problem running across them. A few years ago I saw a photo of Eden Pastora, then the leader of the Nicaraguan contras, in a national magazine. Pastora was seated in front of an amateur transceiver with which, the accompanying story went, he communicated with his troops and argued with radio officers in the regular Nicaraguan army!

Miscellaneous Unusual Signals

There are many other signals heard on SW which are of mysterious origin and purpose. Some of those heard when this book was written are listed in table 10-2; it's certain other mysteries will be on the air by the time you read this.

Some stations do nothing more than transmit various "beeps," "dashes," "pulses," or "ticks" at fixed repetition rates of typically 60 to 90 per second. When first heard, these stations sound similar to time signal transmissions. However, these stations do not identify, nor are they listed in any standard frequency lists such as those of the International Telecommunications Union. Sometimes these signals are interrupted by data bursts or coded groups; it may be these are "markers" of some sort to hold a channel between messages.

Several stations can be heard which transmit what seem to be patterns of dots and dashes using "noise bursts." The effect is similar to trying to send Morse code by hissing it, although the combinations of dots and dashes usually form no letter found in any version of the Morse code. These stations have become known as "raspers" because of their rough, harsh sound. These signals have been around since the 1970s and are rumored to be connected with U.S. military operations, particularly the U.S. Navy. It is apparent that most (if not all) of these signals must originate within the United States due to their signal strength and reception on frequencies unlikely to support propagation over more than a few hundreds of miles. Raspers tend to operate on relatively stable frequencies. However, the patterns of dots and dashes can change within a few hours on the same frequency. As of yet, no one has yet come up with a good explanation of why the patterns change or what the patterns signify.

Not all CW activity you can hear will be maritime stations. There are several CW nets using unusual call signs, such as "OPR" and "Z4U," which can be heard exchanging letter groups with each other. These stations often use hand-sent CW (often with old-style "straight" keys instead of electronic keying devices) and the sending can be very sloppy. On the other hand, some of the nets will have a very professional "sound" with good CW technique and formal check-in/check-out procedures, including the use of military "Z" codes. Strange CW markers have also been heard. Some of these are CQ or VVV markers sent by stations with unusual call signs. Others are seemingly random groups of letters and numbers sent repeatedly.

Some of the more professional nets have been heard on frequencies known to be used by the U.S. Navy, and it's widely suspected these are Navy stations conducting exercises using tactical call signs. The "sloppy" nets have several of the same characteristics used by Cuban civilian maritime radio operators (particularly the horrible CW sending), and it is believed these nets are Cuban in origin, possibly for communications with guerilla forces in Latin America.

The key to hearing all stations in table 10-2 is again patience. Careful tuning outside the "normal" broadcasting and amateur bands will reveal all sorts of surprises. Membership in a SWL club that covers such activity, such as the Association of Clandestine Radio Enthusiasts, is also helpful in determining patterns of activity based upon the receptions of many SWLs.

CHAPTER ELEVEN

The Hobby of Shortwave Listening

IN THIS BOOK, we've sometimes referred to shortwave listening as a hobby. This might have seemed a little unusual to you (does anyone ever say their hobby is "watching television"?), but it does reflect the fact that SWLing often takes a larger commitment of effort and knowledge than listening to your local AM and FM stations.

Shortwave listening became a hobby initially because of the specialized and esoteric nature of SW reception. Until a few years ago, the barriers to understanding and using SW radio were so high that all but the most dedicated individuals were soon discouraged. Today, that's no longer the case and SW radios are no more difficult to use than other consumer electronics devices. Tuning in major international broadcasters such as the BBC or Radio Moscow is as easy as listening to your favorite AM or FM station, and most people who listen to such stations wouldn't say it's their "hobby." So when does SWLing become a hobby?

For me, it happened when I started to get more "systematic" in my approach to listening. I started trying to find new stations in different countries, even if it meant having to identify stations broadcasting in foreign languages. I also began to write those stations and collect the cards and letters they sent back. To keep up with latest happenings, I joined a couple of clubs for SWLs. I got a ham radio license so I could talk to faraway places as well as listen to them. Before I knew it, I was hooked. I was a "real SWL," and it was a hobby for me.

This might not happen to you. You might be content to listen to a few favorite stations, and your only communications with those stations might be letters to get on their mailing lists for program schedules. But I have a feeling that most readers of this book won't have immunity to the "SWLing disease." You'll find yourself anxiously checking your mailbox to see if that long-awaited QSL card has finally arrived. You'll starting memorizing the phrase "This is radio station......" in Indonesian and Arabic. And you'll finding yourself crawling out of bed at 4 a.m. so you can have a chance to hear some obscure station in southeast Asia.

When such things happen, you're a "real SWL."

If you're fated to become a real SWL, you might as well do it right! In this chapter, we'll look at some of the things that make up the hobby of SWLing.

Reception Reports and QSLs

A QSL card or letter is basically mail from a radio station that says "yeah, you heard us"—nothing more, nothing less. I've given up trying to explain why some SWLs (such as myself) will repeatedly try to extract a QSL from a station that has repeatedly refused to answer earlier reports. I think part of the answer lies in the thrill of the hunt. Getting a QSL from a rarely heard station gives the same thrill as finding a rare baseball card, coin, or item of Rin-Tin-Tin memorabilia (I also collect that). Like all collecting efforts, it's difficult to stop until you've got a "complete set." Since new stations are constantly coming on the air, however, you'll always be trying in vain to complete your set!

Entire books could be written on reporting reception to collect QSLs (in fact, one excellent book has been written on the subject). Here, we'll discuss techniques adequate for report-

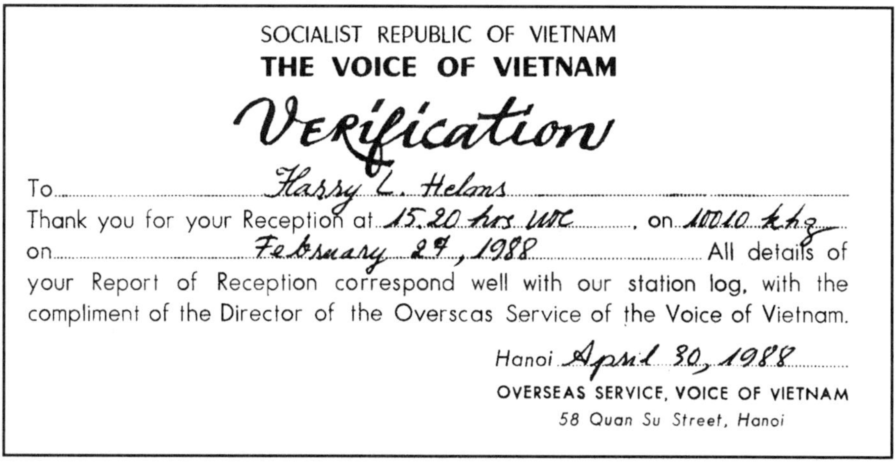

FIGURE 11-1

SOCIALIST REPUBLIC OF VIETNAM
THE VOICE OF VIETNAM

Verification

To: Harry L. Helms
Thank you for your Reception at 15.20 hrs UTC, on 10010 kHz
on February 27, 1988. All details of your Report of Reception correspond well with our station log, with the compliment of the Director of the Overseas Service of the Voice of Vietnam.

Hanoi April 30, 1988
OVERSEAS SERVICE, VOICE OF VIETNAM
58 Quan Su Street, Hanoi

Plain and simple, but it serves it purpose—a QSL card from the Voice of Vietnam in Hanoi.

ing to international broadcasters as well as AM, FM, and TV stations.

Reporting reception for a QSL is a transaction between you and the station you heard, with both of you having something the other wants. You have observations and comments about what you heard, while the station has a card or letter to send you. What you have to do is give something of value to the station that hopefully will make them feel obligated to reply with a QSL card or letter.

Not all SWLs collect QSLs, and by the same token not every station cares to send them out. Many international broadcasters have cut back on sending QSLs, and several persons connected with various international broadcasters—and even some SWLs—have been critical of the entire practice as a waste of limited station resources. However, most international broadcasters have discovered what American and Canadian broadcasters have known for years—some sort of "promotion," whether contests or sending QSLs, is usually necessary to build audience response and participation. On the other hand, many

AM, FM, TV, and domestic SW broadcasters couldn't care less if they're heard outside their intended coverage area. And many utility stations aren't pleased a bit to discover that someone other than their intended recipients may have listened to one of their transmissions.

Many SWLs get very creative in their responses to these situations. One of the most common is for the listener to prepare a QSL card for the station, with all the data filled in, which a station official can then sign and return to the SWL. Other QSL collectors enclose souvenirs, such as picture postcards, decals, or postage stamps, with their reports. While reports to international broadcasters can be in English (or any of the other languages they broadcast in), most foreign domestic SW broadcasters and utility stations in nations where the dominant language isn't English will have no one who understands English; reports to such stations will have to be written in the station's language, often using forms available from various SWL clubs. To ensure that their reception report arrives at a station, and to give it the aura of importance, some listeners use registered mail if their first couple of reports go unanswered. You'll need patience to extract a QSL from many stations; some hardcore QSL collectors have sent more than a dozen reports to a single station over several years in hopes that someone will eventually verify one of their reports.

What are the basic elements of any reception report? Here they are:

- the frequency you heard the station on
- the month, day, and year you heard the station
- the time your reception began and ended
- the quality of the station's signal, including signal strength and QRM
- a brief description of your radio and antenna
- if the station is an international broadcaster, some comments and reactions to their programs

- a *request* for the station to verify your report if it is correct
- enough details of what you heard to prove that you actually heard the station

If you're using a receiver with a direct frequency readout, report the frequency it indicates; otherwise, you'll have to rely on announcements by the station. Sometimes you'll note that the announced frequency does not match the received frequency, as when Radio Moscow announces a rounded frequency such as "9.53 MHz" when they're actually transmitting on 9535 kHz. The best policy in such cases is to report both the announced and actual frequencies.

The date and time items are interrelated. If you're reporting to an international broadcaster or utility station, the time should *always* be given in UTC even if the station itself gives times in the local time zones of its intended target areas. If you're reporting to a domestic station, the best bet is to report in the local time zone where the station is located. Thus, if you live in the Mountain time zone and hear a BCB station in the Eastern, you'll report using Eastern time. Moreover, the date you use in your report depends upon the time used in your report. If you hear a European station during the evening in North America, it's the next day in UTC, and that's the date you should use if you report in UTC.

Indicate the beginning and ending times of your reception. How long should you listen? If you're reporting to a major international broadcaster, a half-hour is a good minimum. In other cases, the reception might be unavoidably brief, as in sporadic-E or other short-lived propagation openings. Most utility transmissions are short by their very nature (and there's no use in listening long to a repeating marker transmission). And there's a limit to what you can report if a station signs off soon after you first tune in.

For years, many SWLs have used the "SINPO" code to indicate how well a station was received. "SINPO" is an acronym for *S*ignal strength, *I*nterference, *N*oise, *P*ropagation,

FIGURE 11-2

In addition to QSL cards, many stations—such as Radio Korea—also send out pennants to reporters.

and Overall quality. The scale used ranges from 1 to 5, with 5 denoting the best possible condition (loudest signal, no interference, no noise, excellent propagation, and superb overall quality) and 1 representing a condition so poor it makes the signal unusable or unlistenable. Although it's a convenient shorthand (and I've used it myself), I don't think a lot of the SINPO code. Its biggest problem is that fallacy found whenever an "objective" number is assigned to an inherently subjective evaluation. What I might call a "SINPO 44544" signal another SWL might call "SINPO 53433." And neither of us could be proven correct or incorrect! This also shows up in some SWL reports with clearly illogical SINPO ratings, such as "SINPO

32324." Despite these problems, some international broadcasters and SWL clubs still promote the use of SINPO.

I think a clearly written description of the signal quality is just as "objective" and much more useful. A description such as "your signals were of good to very good strength, with only slight regular fading, but with moderate to heavy interference at times from the BBC relay at Antigua on 6185 kHz" conveys far more useful information than "SINPO 43543." Moreover, SINPO doesn't take into account the dynamic nature of SW reception; a written description lets you indicate the amount of interference from the BBC varied and the rate of fading and other propagation disturbances. You'll notice the written description didn't mention noise. In most cases, noise is generated in the SWL's local environment; if you have noise from a power transformer down the street, there's nothing the station can do about it. If, however, there had been high atmospheric noise due to solar activity or lightning QRN, then that would have been an item to include in the written description.

Don't use SINPO when reporting to domestic broadcasters (including AM, FM, and TV broadcasters) or utility stations—they won't have a clue as to what it means!

If a station is interested that you heard them, they'll be interested in your receiver and antenna as well. You don't have to list and describe every single item of radio equipment you have, but do mention the make and model of the receiver you use. It's often helpful to add a brief descriptive phrase, such as "portable shortwave radio" or "communications receiver" in case the station personnel aren't familiar with a particular model. If you had to use a special feature or technique to receive the station, such as preamplifier or ECSSB, mention this. Describe your antenna as a longwire, random wire, etc., and give its dimensions (length and height above ground) in meters and centimeters instead of inches and feet.

Broadcasters like comments about their programming. Even if you don't understand the language a program was in (as with domestic SW stations), you could remark that you enjoyed the music or are interested in learning the language the station broadcasts in. Major international broadcasters are becoming increasingly interested in comments on their programming, judging from remarks of personnel at various international stations who have been quoted in SWL publications. Some of these comments have seemed to imply that SWLs have favorable remarks and useful suggestions to make about programs, but don't in their haste to secure a QSL. Others, such as the current management of Radio Finland, have bluntly said they don't want to divert funds that could be used for programming production to answering reception reports and sending QSLs. (It never seems to occur to such station personnel that their programming might bore most listeners stiff, and indeed the *only* rational reason for listening to the station would be to secure a QSL!)

Remarks about programming are an essential part of any report to an international broadcaster, as long as those remarks are *honest* and *candid*. Some SWLs are tempted to say nice things about a station's programs just to improve their chances of getting a QSL or other goodie such as a pennant, but doing so only perpetuates mediocre programming practices. Several stations say they want "constructive" criticism of their programming, and by "constructive" they often mean only favorable remarks or at least offsetting any critical remarks by an equal number of favorable remarks. I suggest that you interpret "constructive" as "honest." This doesn't mean you have to go out of your way to be negative, but it does mean you should give your genuine reactions to a station's programs. The improvements in transmitting facilities and SW radios have meant that the "hardware" is no longer a problem in shortwave broadcasting. However, the "software"—the programming—often still is. Too many countries have spent large sums on broadcasting equipment that will let you hear them, without improving the programs enough so that people will want to listen. Programming

FIGURE 11-3

A single report to a station will likely be enough to get you on their mailing list for program schedules for the next several years.

all too often sounds as if it was produced by committees using formulas. Too many programs have no relevance to listeners in target areas, and the broadcasters don't seem to realize that what works in their home country or culture doesn't necessarily work with overseas listeners. The only way this situation will change is if SWLs bring it to the attention of those managing international broadcasters. Praise what you like—but don't be reluctant to say what you don't like or to ask for something different.

It's also important to *request* that a station verify your report instead of demanding that it do so. One reason is simple courtesy to the station personnel who read letters from listeners. (Think about it for a second—you have a letter from a total stranger in a faraway country who wants you to do a favor for him or her. Would you be more likely to do so if it were a friendly request for your help or a blunt, direct order?) Another is that some stations will only send a QSL upon request. A pet phrase I like to use in my reports is "If this reception report is correct and of use to your station's staff, I would greatly appreciate receiving a card or letter confirming that I indeed heard your station at the time and on the date and frequency I noted in this letter." If you're reporting to a station such as a utility, you may find it's effective to explain why you want a QSL from that station, such as "My hobby is trying to hear as many different shortwave (or AM, FM, and so on) stations as I can, and to collect cards and letters from these stations confirming that I did indeed hear them. Thus, a card or letter from your station would be a valued addition to my collection."

The major part of your reception report will likely be devoted to details and information to prove that you indeed heard the station. Admittedly, many stations do not fully check reports, and it's not unheard of for a listener to receive a "QSL" in response to a request for a program schedule. But some stations do carefully check reports, and some stations retain reports in their files or release them to local SWL clubs (for prospective members, and the like). Some SWLs have unfortunately developed reputations as "reception fakers," persons whose impressive QSL collections are based upon fraudulent reports. It's often possible to "fake" receptions based upon material appearing in SWLing publications, but the damage to a SWL's reputation can be (and usually is) lasting if one of the fraudulent reports circulates within the SWL hobby. (One well-known DXer, now deceased, had his reputation permanently tarnished by a reception report to a New Zealand BCB station in which the "details" consisted of nothing more than the station identification announcement taken from the *World Radio Television Handbook*; unfortunately for this DXer, the announcement had changed since publication of the last edition of the *Handbook*.)

The usual format is to list each item you are quoting by the time at which you heard it. The best possible material to prove your reception are items which you could not have possibly known about without actually having heard the station. For example, did the station sign off or on at a different time than scheduled? Did the station use a new frequency or perhaps drift off its assigned one? Were there unusual technical difficulties? Was there a sudden, unexpected change in normally scheduled programming?

The next best items, particularly if you're reporting normally scheduled programs, are things such as the names of the announcers, titles of musical selections played, items given in a newscast, topics of any commentaries and news analyses, names of listeners whose letters were read over the air, and the like. Advertisements are excellent items to quote in reporting reception of commercial stations. The least convincing items are those which could be gathered from program schedules or SWLing publications. Remember, the burden is on you to prove that you heard the station, and using items that are "public knowledge" or vague doesn't help your cause.

The tendency of some stations to send out QSLs indiscriminately means you shouldn't rely on a QSL to prove you heard the station if you're not positive yourself that you heard it. Some SWLs send out "tentative" reports which, in effect, ask the station "did I really hear you?" This technique is acceptable if you have strong reason to believe that you did indeed hear the station, but shouldn't be relied upon if there are two or more possible stations you could've heard and you're not sure which one it was.

The entire matter of whether a QSL "proves" reception is often hotly debated in the SWL hobby, particularly among SWLs and DXers who engage in the more competitive aspects of the hobby. The strong consensus is that a QSL by itself proves nothing; whether a SWL's report of hearing a rare station is believed will depend more on that SWL's reputation as an accurate, reliable, and honest reporter of what he or she heard than upon any QSL.

If you report to foreign stations, the postage bill will quickly add up. Airmail is more expensive than seamail, but it's the only viable way to make sure your report arrives in time to have any value to a station. To keep your postage bill down, use onion-skin paper and envelopes for your reports. If you don't need to enclose return postage or other items with a report, an *air letter sheet* or *aerogramme* can save money. Available from any post office, these are sheets of paper with preprinted airmail postage that's slightly less than that for an airmail letter. You write your letter on one side of the sheet, fold it into the shape of an envelope, and seal it using the gummed flaps provided. There's enough room for a complete report on an aerogramme sheet. If you write to major international broadcasters, aerogrammes are the way to go.

Major international broadcasters will QSL or send out program schedules without return postage. (Some require return postage if you want the reply sent airmail.) Domestic and private SW stations, as well as international broadcasters operated by less-affluent nations, do appreciate (and often require) return postage for QSLs. The traditional method of doing so has been to send one or more *international reply coupons* (IRCs), which are currently available from larger post offices.

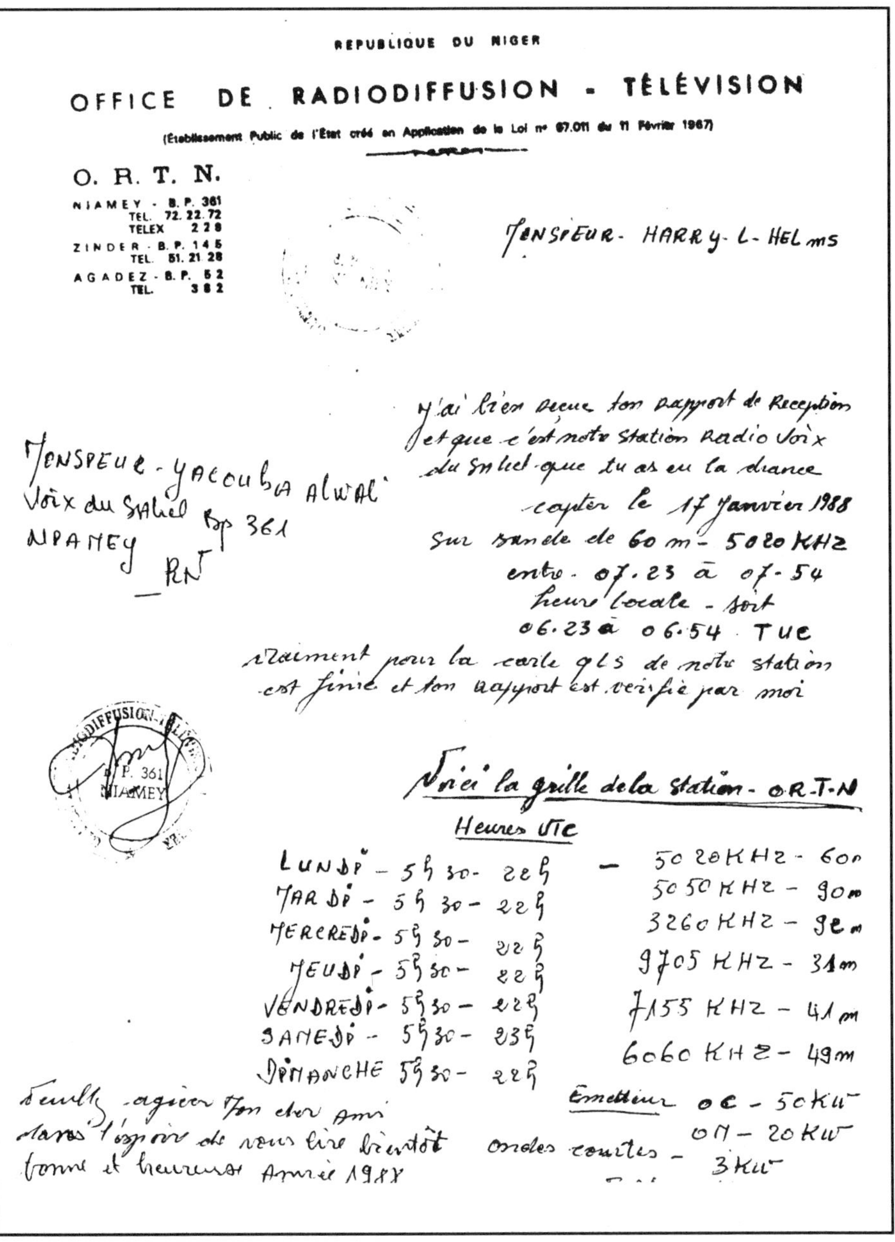

REPUBLIQUE DU NIGER

OFFICE DE RADIODIFFUSION - TÉLÉVISION

(Établissement Public de l'État créé en Application de la Loi n° 67.011 du 11 Février 1967)

O. R. T. N.

NIAMEY - B.P. 361
TEL. 72.22.72
TELEX 228
ZINDER - B.P. 145
TEL. 51.21.28
AGADEZ - B.P. 52
TEL. 382

Monsieur - Harry - L - Helms

Monsieur - Yacouba Alwali
Voix du Sahel Bp 361
Niamey
_RN

J'ai bien reçu ton rapport de Réception et que c'est notre station Radio Voix du Sahel que tu as eu la chance capter le 17 janvier 1988 sur bande de 60 m - 5020 KHZ entre 07.23 à 07-54 heure locale - soit 06.23 à 06.54 TUC

vraiment pour la carte QLS de notre station est finie et ton rapport est vérifié par moi

B.P. 361 NIAMEY

Voici la grille de la station - O.R.T.N

Heures UTC

LUNDI	5h30 - 22h	5020 KHZ - 60m
MARDI	5h30 - 22h	5050 KHZ - 90m
MERCREDI	5h30 - 22h	3260 KHZ - 90m
JEUDI	5h30 - 22h	9705 KHZ - 31m
VENDREDI	5h30 - 22h	7155 KHZ - 41m
SAMEDI	5h30 - 23h	6060 KHZ - 49m
DIMANCHE	5h30 - 22h	

Émetteur OC - 50KW
OM - 20KW
ondes courtes - 3KW

Veuillez agréer mon cher ami dans l'espoir de vous lire bientôt bonne et heureuse Année 1988

Amid all the preprinted QSL cards, you'll sometimes receive a real gem such as this handwritten letter from Niger.

An IRC can be exchanged in virtually any country of the world for postage stamps to pay for a *surface* letter to a foreign country; thus, you'll need to send two or three IRCs for an airmail reply.

IRCs aren't usually the best way to go, however. In many nations, IRCs may not be understood by local postal officials and it's not unusual for them to refuse to redeem IRCs. Moreover, the nearest post office that redeems IRCs might be a long way from the station; it might take an entire day for a trip to and from such a post office. And IRCs are expensive; at the time this book was written, they were almost a dollar each and the price was expected to rise. More useful in many cases are mint stamps of the country. Such stamps can be obtained from stamp dealers or one of the various "DX stamp services" that sell to hams and SWLs. Mint stamps are invariably cheaper than IRCs, even after the stamp dealers' profits are factored in. Finally, a growing number of SWLs are finding that a single U.S. dollar with a report is the most effective way to get a reply. This amount is almost always enough to pay for return airmail postage and usually leaves the station a little bit left over, a minor but thoughtful gesture for many financially-strapped stations.

Another method to increase the effectiveness of your reception reports is to make them as legible as possible by using a computer printer or typewriter, or by printing carefully. Most handwriting is difficult enough to read already, and imagine the problems faced by station personnel whose first language may not be English!

A growing number of SWLs are "QSLing" their receptions by tape recording their receptions. Some stations appreciate receiving cassette tapes of their signals, and often return those tapes with music or programming recorded on them. Other SWLs just keep the tapes for their own pleasure, and many have taped libraries of hundreds of their receptions. A tape recording of a reception along with a written QSL from a station is an unbeatable combination for many SWLs!

Record Keeping and Awards

Many SWLs and DXers have a habit of keeping track of what they hear in a written record known as a *log*. This is a bit like a diary, with the exact format and content up to the individual. One common use for a log is to keep track of the different stations and countries one hears and verifies, and some DXers can instantly rattle off the exact numbers they've achieved. These are known as "totals."

Logs come in all shapes, sizes, and types. Some SWLs like to use preprinted logging sheets and forms, available from SW equipment dealers, while others use ruled notebooks like those used by students. The amount of information included also varies with personal preference. Some are content to just record the time, frequency, and name or call sign of a station while others record virtually every detail of each reception.

Some SWL clubs encourage this sort of record keeping, as they offer awards and certificates for collecting QSLs from a specified number of countries or other DX targets, such as the USSR republics or Nigerian regional stations. A written log is important in such cases for determining which countries or stations are still needed or which haven't answered reception reports.

Even if you don't care about awards or keeping track of how many stations you've heard, a log can still be useful. For example, it can be a good tool for determining reception patterns from different parts of the world at different seasons of the year. If the K-index and solar flux values are noted in the log, it can serve as a guide to what can be expected under similar conditions. Other SWLs make their log into a diary of sorts, complete with observations and remarks about the general condition of the SW bands, results of comparisons between different antennas and equipment, and so forth. If you get interested in following SW puzzles such as numbers stations, some sort of written record will be essential to detect patterns of activity.

Of course, there's no requirement that you keep a log or any other written record of your listening experiences. The choice is yours.

SWL "Call Signs" and Identifiers

Back in the late 1950s, the editors of the original *Popular Electronics* magazine (not related to the current magazine bearing that name) came up with the idea of assigning "identifiers," similar to amateur call signs, to SWLs who desired them. While a new idea in America, this practice had been going on for some time in Europe (particularly Soviet bloc nations) and was intended for those SWLs who aspired to become hams; such SWLs were issued identifiers by the national amateur radio organizations so they could feel like "one of the gang" before actually getting licensed. The editors of *Popular Electronics* decided to begin their identifiers with the prefix "WPE" followed by a digit (corresponding to the amateur radio call sign district) and three more letters. Not only did this arrangement avoid any conflict with actual call sign assignments, it was a good advertisement for the magazine.

Thousands of "WPE monitoring certificates" were issued by *Popular Electronics* and soon SWLs began to include their WPE "call letters" in correspondence to the magazine, stations, and other SWLs. Not that these calls and certificates were handed out to just anyone; you had to have received five QSLs, with at least one from a station outside the United States—and don't forget the 25¢ in coins (no stamps, please!). Some SWLs even had their own "QSL" cards printed with their calls, and were soon busy swapping cards with other SWLs and including their calls in reception reports. By the mid-1960s, it seemed as if every self-respecting SWL had a WPE "call" and "QSL" cards to boot.

However, by the late 1960s new editorial management at *Popular Electronics* decided to cut back on the amount of SWL-related material in the magazine, including the discontinuance of the monthly SWL column and the entire WPE program. The magazine's SWL column editor took over what remained of the WPE program, and continued issuing identifiers beginning with the "WDX" prefix. Since the WPE/WDX program was mainly intended for SWLs listening on 30 MHz and below, another registration program was started by a company specializing in frequency guides and directories for scanner listeners and VHF/UHF buffs. Prefixes in this program begin with "K" followed by the post office two-letter abbreviation for the listener's home state. But deprived of the driving engine of *Popular Electronics*, the entire business of SWL calls and card swapping soon faded. Today, it is still practiced but by only a fraction of the numbers who participated when the craze was at its peak. If you do run across what appears to be a call sign beginning with "WDX2" or "KFL4," however, you'll know it's a SWL or scanner listener and not an actual transmitting station.

Many SWLs who were around during the height of the WPE mania have fond memories of that period, and one club devoted to WPE calls and other 1960s SWLing nostalgia was formed. Indeed, there was something special about SW radio in that era before communications satellites and live TV from the moon, and SWLs felt part of a select, secret few. (And, yes, I was proud to be known as WPE4HKE back in that period!)

SWL Clubs

SWL clubs are nonprofit associations of SWLs who band together to exchange information and tips regarding their particular listening interests. All labor is voluntary and unpaid, and the work load of publishing a monthly (or more frequent) club bulletin and administering club affairs is often a heavy burden. For most club officers and editors, the only reward they derive is the satisfaction of knowing they do a good and vital job or, if they actually publish the bulletin, in getting a copy of it before anybody else. The various columns and features in club bulletins are put together by unpaid editors, who donate their time and efforts for the benefit of the rest of the club. All clubs charge dues for membership, but these are almost exclusively consumed by the expenses of printing and mailing the bulletin as well as club management and administration expenses. (There is an almost perfect correlation between increases in postage rates and increases in club membership dues.) If one wants to get rich, starting and running a SWL club is not the way to do it!

Most club bulletins today are printed by using the "offset" technique; this results in a bulletin printed as a booklet which can be easily mailed and read. All material is prepared using computer printers or typewriters; illustrations of QSL cards and photos can also be run. Some smaller clubs and groups still use mimeographs or similar techniques to produce their bulletins, but these are becoming rarer.

Some clubs are democratically governed, with elections of club officers to manage club affairs and develop policies; the publication of the bulletin is in the hands of a separate publications group. Most clubs, however, are managed by the same group that publishes the bulletin. The principle in this case is that those who do the work, which is unpaid, should have the final say in running the club. The lack of a "democratic" structure doesn't mean such clubs are necessarily dictatorships or totalitarian. Almost all such clubs are very responsive to member needs and concerns. It's also true that in most cases there's no real alternative to letting the publishers control the club. Enormous time and effort is needed to publish the bulletin and manage a club, and without someone to do so there's no club. In fact, the biggest reason why SWL and DX clubs go out of existence is an inability to find someone willing to take over responsibility for publishing the bulletin when the existing publisher resigns. The real test of any form of club government and management is how well a club serves its membership.

FIGURE 11-5

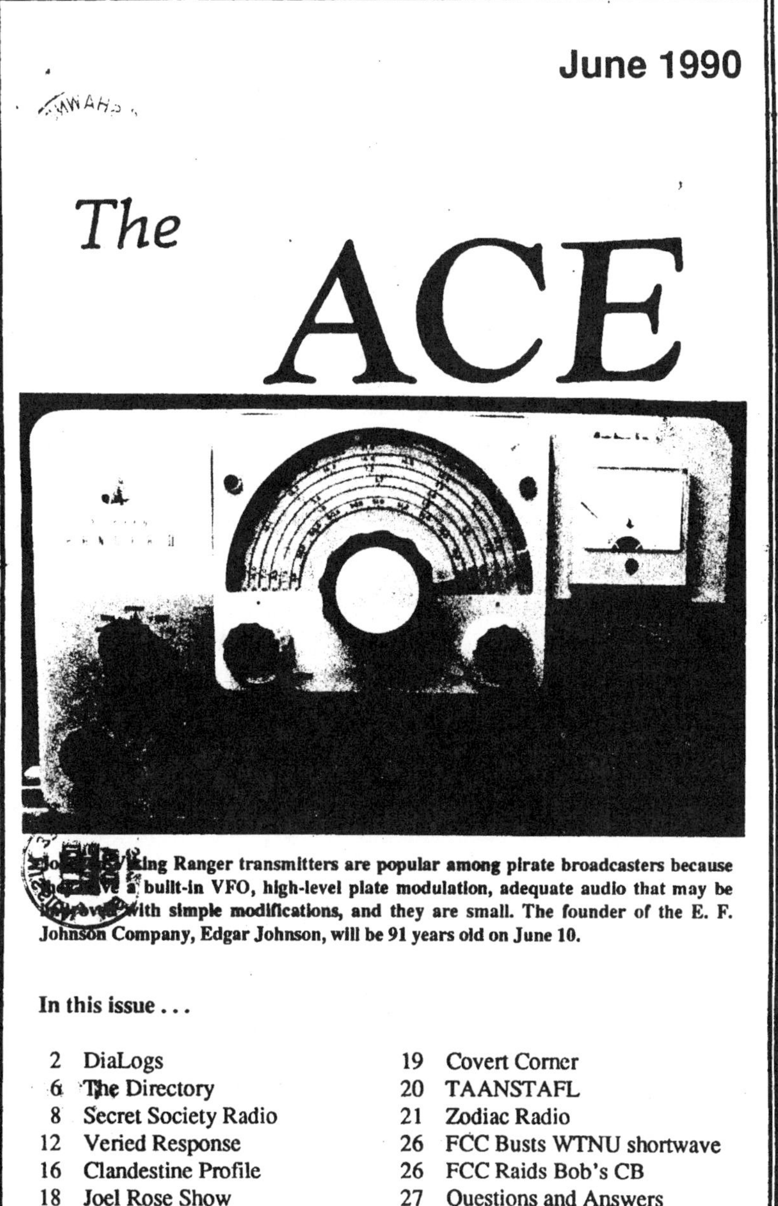

June 1990

The ACE

Johnson Viking Ranger transmitters are popular among pirate broadcasters because they have a built-in VFO, high-level plate modulation, adequate audio that may be improved with simple modifications, and they are small. The founder of the E. F. Johnson Company, Edgar Johnson, will be 91 years old on June 10.

In this issue . . .

The Association of Clandestine Radio Enthusiasts sends out its monthly bulletin, "The ACE," to its members worldwide.

Most SWLs couldn't care less how a club is controlled so long as a club's bulletin and other services enhance their listening!

Clubs vary by the scope of their coverage and by their emphasis on SWLing and DXing. Some clubs are "all band," meaning they cover everything from AM DXing to SWLing to FM/TV DX. Other clubs specialize in segments of the hobby such as AM DXing, pirate and clandestine radio, or LW reception. Some clubs place more emphasis upon "DXing," with its pursuit of new countries, stations, and QSLs, while other clubs are more oriented toward casual SWLing and news of schedule changes by major international broadcasters. Most all band clubs today try to strike a balance between the two interests, with the specialty clubs being more oriented toward DXing.

Which SWL club (or clubs) should you join? There's no easy answer to that, since a lot will depend on your own preferences and interests. Moreover, the strengths and weaknesses of clubs change over time, depending on the editors and members involved. It's not unknown for a strong, vital club to have its bulletin rapidly deteriorate and its membership drop following the resignation of key editors or club officers. By the same token, a moribund club can be revitalized by new editorial talent.

Some clubs also have a tendency to get involved in areas that have little, if any, relevance to SWLing and DXing or disintegrate into soapboxes for the political and social opinions of various editors and club officials. For others, being a club official or editor provides their first taste of "power" and "status," resulting in egos running amok. Childish disputes can erupt and reverberate for years. (I remember the case of one DXer who was expelled when he was a teenager from a club. Six years later—then a college senior—this same DXer sent letters of acceptance he had received from various law schools to his former "enemies," apparently in an attempt to show them how "wrong" they had been!)

Other SWLs seem to enjoy stirring up controversies with clubs; often, the motivation seems to be their need for attention at any cost and a lack of anything better to do with their time or energy. Such controversies are rarely over substantive matters, but are purely personality clashes. Fortunately, you don't have to get involved with such nonsense—or support any club that allows such irrelevance to take up space which could otherwise be filled with SWLing and DXing news.

The only way to judge whether or not a club might be for you is to examine a sample copy of its bulletin. Fortunately, almost every club will be willing to send you a sample bulletin and membership information. Since clubs are nonprofit ventures, it's best to enclose a dollar to cover the costs of the sample bulletin and mailing. A list of addresses for major SWL clubs is included in the appendix for this book.

Once you find clubs that agree with your needs and interests, you'll find them valuable adjuncts to your listening. One aspect many like is the opportunity for contact with other SWLs. It's always interesting to see the kind of reception others in your general area or those using similar equipment are able to achieve. Some clubs allow members space for their opinions and observations, and this can be an important part of getting to know other members better. (Unfortunately, it can also lead to some of the problems cited earlier.) Most clubs also permit members to buy and sell receivers and other equipment through free ads in the bulletin. Some clubs also sponsor regional chapters and get-togethers.

On-Line Bulletin Boards and Conferences

SWL-to-SWL contact has entered the computer age through the establishment of various "bulletin boards" which can be accessed by those with a personal computer and modem for telephone line communications. These bulletin boards allow tips and information to be exchanged in a matter of minutes with other listeners. Currently, the Association of Clandestine Radio Enthusiasts sponsors an "all band" board for its members and all interested SWLs; other boards are operated by various clubs and individuals for more specialized listening interests.

At the time this book was written, regular on-line teleconferences for SWLs and DXers were being held on the Portal Telecommunications System and GEnie. Portal is home to "Los Numeros On-Line," hosted by well-known numbers station expert Havana Moon. In these sessions, SWLs from across the country are linked together to chase numbers stations and pirate broadcasters.

Linking SWLs by computer is in its infancy. Computers will doubtlessly become a major way for SWLs to communicate and share information in the future.

APPENDIX ONE

Call Sign Allocations of the World

AAA–ALZ	United States
AMA–AOZ	Spain
APA–ASZ	Pakistan
ATA–AWZ	India
AXA–AXZ	Australia
A2A–A2Z	Botswana
A3A–A3Z	Tonga
A4A–A4Z	Oman
A5A–A5Z	Bhutan
A6A–A6Z	United Arab Emirates
A7A–A7Z	Qatar
A9A–A9Z	Baharain
BAA-BZZ	China (BVA–BVZ used by Taiwan)
CAA–CEZ	Chile
CFA–CKZ	Canada
CLA–CMZ	Cuba
CPA–CPZ	Bolivia
CQA–CUZ	Portugal and its territories
CVA–CXZ	Canada
C2A–C2Z	Nauru
C3A–C3Z	Andorra
C4A–C4Z	Cyprus
C5A–C5Z	The Gambia
C6A–C6Z	Bahamas
C8A–C9Z	Mozambique
DAA–DTZ	Germany
DUA–DZZ	Philippines
D2A–D3Z	Angola
D4A–D4Z	Cape Verde Islands
D6A–D6Z	Comoros
EAA–EHZ	Spain and its territories
EIA–EHZ	Ireland
EKA–EKZ	USSR
ELA–ELZ	Liberia
EMA–EOZ	USSR
EPA–EQZ	Iran
ERA–ERZ	USSR
ESA–ESZ	Estonian SSR
ETA–ETZ	Ethiopia
EUA–EWZ	Byelorussian SSR
EXA–EZZ	USSR
FAA–FZZ	France and its territories
GAA–GZZ	United Kingdom
HAA–HAZ	Hungary
HBA–HBZ	Switzerland
HCA–HDZ	Ecuador
HEA–HEZ	Switzerland
HFA–HFZ	Poland
HGA–HGZ	Hungary
HHA–HHZ	Haiti
HIA–HIZ	Dominican Republic
HJA–HKZ	Colombia
HLA–HMZ	South Korea
HNA–HNZ	Iraq
HOA–HPZ	Panama
HQA–HRZ	Honduras
HSA–HSZ	Thailand
HTA–HTZ	Nicaragua
HUA–HUZ	El Salvador
HVA–HVZ	Vatican City
HWA–HWZ	France and its territories
HZA–HZZ	Saudi Arabia
IAA–IZZ	Italy and its administered territories
JAA–JSZ	Japan
JTA–JVZ	Mongolia
JWA–JXZ	Norway
JYA–JYZ	Jordan
J2A–J2Z	Djibouti
J3A–J3Z	Grenada
J5A–J5Z	Guinea-Bissau
J6A–J6Z	St. Lucia
J7A–J7Z	Dominica
J8A–J8Z	St. Vincent
KAA–KZZ	United States
LAA–LNZ	Norway
LOA–LWZ	Argentina
LXA–LXZ	Luxembourg
LYA–LZZ	Lithuanian SSR
L2A–L9Z	Argentina
MAA–MZZ	United Kingdom

NAA–NZZ	United States
OAA–OCZ	Peru
ODA–ODZ	Lebanon
OEA–OEZ	Austria
OFA–OJZ	Finland
OKA–OMZ	Czechoslovakia
ONA–OTZ	Belgium
OUA–OZZ	Denmark and its territories (including Greenland)
PAA–PIZ	Netherlands
PJA–PJZ	Netherlands Antilles
PKA–POZ	Indonesia
PPA–PYZ	Brazil
PZA–PZZ	Surinam
P2A–P2Z	Papua New Guinea
P5A–P5Z	North Korea
QAA–QZZ	Reserved for international "Q" signals
RAA–RZZ	USSR
SAA–SMZ	Sweden
SNA–SRZ	Poland
SSA–SSM	Egypt
SSN–STZ	Sudan
SUA–SUZ	Egypt
SVA–SVZ	Greece
S2A–S3Z	Bangladesh
S6A–S6Z	Singapore
S7A–S7Z	Seychelles
S8A–S8Z	Transkei district of South Africa
S9A–S9Z	Sao Tome e Principe
TAA–TCZ	Turkey
TDA–TDZ	Guatemala
TEA–TEZ	Costa Rica
TFA–TFZ	Iceland
TGA–TGZ	Guatemala
THA–THZ	France and its territories
TIA–TIZ	Costa Rica
TJA–TJZ	Cameroon
TKA-TKZ	France and its territories
TLA–TLZ	Central African Republic
TMA–TMZ	France and its territories
TNA–TNZ	Congo
TOA–TQZ	France
TRA—TRZ	Gabon
TSA–TSZ	Tunisia
TYA–TYZ	Benin
TZA–TZZ	Mali
T2A–T2Z	Tuvalu
T3A–T3Z	Kiribati
UAA–UZZ	USSR
VAA–VGZ	Canada
VHA–VNZ	Australia
VOA–VOZ	Canada
VPA–VSZ	British colonies and administered areas
VTA–VWZ	India
VZA–VZZ	Australia
V2A–V2Z	Antigua and Barbuda
V3A–V3Z	Belize
V4A–V4Z	St. Christopher
V8A–V8Z	Brunei
WAA–WZZ	United States
XAA–XIZ	Mexico
XJA–XOZ	Canada
XPA–XPZ	Denmark and its territories (including Greenland)
XQA–XRZ	Chile
XSA–XSZ	China
XTA–XTZ	Burkina Faso
XUA–XUZ	Cambodia
XVA–XVZ	Vietnam
XWA–XWZ	Laos
XXA–XXZ	Portuguese territories
XYA–XZZ	Burma
YAA–YAZ	Afghanistan
YBA–YHZ	Indonesia
YIA–YIZ	Iraq
YJA–YJZ	New Herbides
YLA–YLZ	Latvian SSR
YMA–YMZ	Turkey
YNA–YNZ	Nicaragua
YOA–YRZ	Romania
YSA–YSZ	El Salvador
YTA–YUZ	Yugoslavia
YVA–YVZ	Venezuela
YZA–YZZ	Yugoslavia
ZAA–ZAZ	Albania
ZBA–ZJZ	British territories
ZKA–ZMZ	New Zealand
ZNA–ZOZ	British territories
ZPA–ZPZ	Paraguay
ZQA–ZQZ	British territories
ZRA–ZUZ	South Africa

ZVA–ZVZ	Brazil
Z2A–Z2Z	Zimbabwe
1AA–1AZ	Unofficially used in disputed territories
2AA–2ZZ	United Kingdom
3AA–3AZ	Monaco
3BA–3BZ	Mauritius
3CA–3CZ	Equitorial Guinea
3DA–3DM	Swaziland
3DN–3DZ	Fiji
3EA–3FZ	Panama
3GA–3GZ	Chile
3HA–3UZ	China
3VA–3VZ	Tunisia
3WA–3WZ	Vietnam
3XA–3XZ	Guinea
3YA–3YZ	Norway
3ZA–3ZZ	Poland
4AA–4CZ	Mexico
4DA–4IZ	Philippines
4JA–4LZ	USSR
4MA–4MZ	Venezuela
4NA–4OZ	Yogoslavia
4PA–4SZ	Sri Lanka
4TA–4TZ	Peru
4UA–4UZ	United Nations
4VA–4VZ	Haiti
4WA–4WZ	Yemen
4XA–4XZ	Israel
4YA–4YZ	International aviation
4ZA–4ZZ	Israel
5AA–5AZ	Libya
5BA–5BZ	Cyprus
5CA–5GZ	Morocco
5HA–5IZ	Tanzania
5JA–5KZ	Colombia
5LA–5MZ	Liberia
5NA–5OZ	Nigeria
5PA–5QZ	Denmark and its territories
5RA–5SZ	Malagasy Republic
5TA–5TZ	Mauritania
5UA–5UZ	Niger
5VA–5VZ	Togo
5WA–5WZ	Western Samoa
5XA–5XZ	Uganda
5YA–5ZZ	Kenya
6AA–6BZ	Egypt
6CA–6CZ	Syria
6DA–6JZ	Mexico
6KA–6NZ	South Korea
6OA–6OZ	Somalia
6PA–6SZ	Pakistan
6TA–6UZ	Sudan
6VA–6WZ	Senegal
6XA–6XZ	Malagasy Republic
6YA–6YZ	Jamaica
6ZA–6ZZ	Liberia
7AA–7IZ	Indonesia
7JA–7NZ	Japan
7OA–7OZ	Yemen
7PA–7PZ	Lesotho
7QA–7QZ	Malawi
7RA–7RZ	Algeria
7SA–7SZ	Sweden
7TA–7YZ	Algeria
7ZA–7ZZ	Saudi Arabia
8AA–8IZ	Indonesia
8JA–8NZ	Japan
8OA–8OZ	Botswana
8PA–8PZ	Barbados
8QA–8QZ	Maldives
8RA–8RZ	Guyana
8SA–8SZ	Sweden
8TA–8YZ	India
8ZA–8ZZ	Saudi Arabia
9AA–9AZ	San Marino
9BA–9DZ	Iran
9EA–9FZ	Ethiopia
9GA–9GZ	Ghana
9HA–9HZ	Malta
9IA–9JZ	Zambia
9KA–9KZ	Kuwait
9LA–9LZ	Sierra Leone
9MA–9MZ	Malaysia
9NA–9NZ	Nepal
9OA–9TZ	Zaire
9UA–9UZ	Burundi
9VA–9VZ	Singapore
9WA–9WZ	Malaysia
9XA–9XZ	Rwanda
9YA–9ZZ	Trinidad and Tobago

APPENDIX TWO

International Phonetic Alphabet

Letter	Phonetic Equivalent	Letter	Phonetic Equivalent
A	Alpha	N	November
B	Bravo	O	Oscar
C	Charlie	P	Papa
D	Delta	Q	Quebec
E	Echo	R	Romeo
F	Foxtrot	S	Sierra
G	Golf	T	Tango
H	Hotel	U	Uniform
I	India	V	Victor
J	Juliet	W	Whiskey
K	Kilo	X	X-ray
L	Lima	Y	Yankee
M	Mike	Z	Zulu

International Morse Code by Sound Chart

Letter	Sound Equivalent	Letter	Sound Equivalent
A	Didah	U	Dididah
B	Dahdididit	V	Didididah
C	Dahdidahdit	W	Didahdah
D	Dahdidit	X	Dahdididah
E	Dit	Y	Dahdidahdah
F	Dididahdit	Z	Dahdahdidit
G	Dahdahdit	1	Didahdahdahdah
H	Dididit	2	Dididahdahdah
I	Didit	3	Didididahdah
J	Didahdahdah	4	Dididididah
K	Dahdidah	5	Dididididit
L	Didahdidit	6	Dahdididit
M	Dahdah	7	Dahdahdididit
N	Dahdit	8	Dahdahdahdidit
O	Dahdahdah	9	Dahdahdahdahdit
P	Didahdahdit	0	Dahdahdahdahdah
Q	Dahdahdidah	.	Didahdidahdidah
R	Didahdit	?	Dididahdahdidit
S	Dididit	,	Dahdahdididahdah
T	Dah		

Q-Signals Used in Morse Code Communication

All signals can be made into questions by following each with a question mark.

Signal	Meaning
QRL	I am busy
QRM	Your transmission has interference
QRN	I am troubled by static
QRO	Increase transmitter power
QRP	Decrease power
QRS	Send more slowly
QRT	Stop transmitting
QRU	I have nothing for you
QRV	I am ready
QRX	Call again
QRZ	You are being called by station —
QSL	I acknowledge receipt
QSO	I can communicate with —
QSX	I am listening on —
QSY	Change your frequency to —
QTH	My location is —

Abbreviations Used in Morse Code Transmissions

ABT	About
AGN	Again
ANT	Antenna
BK	Break
B4	Before
CK	Check
CL	Call
CQ	General call to any station
CUD	Could
CUL	See you later
DX	Distance; distant stations
ES	And
FB	Fine business
GE	Good evening
GM	Good morning
GN	Good night
GND	Ground
HI	Laughter
HR	Here
HV	Have
LID	A poor or careless operator
NR	Number
OM	Old man (a general term used for any male radio operator)
OP	Operator
PSE	Please
PWR	Power
RCVR	Receiver
RX	Receiver
SIG	Signal
SKED	Schedule
SRI	Sorry
TNX	Thanks
UR	Your
VY	Very
WUD	Would
WX	Weather
XCVR	Transceiver
XMTR	Transmitter
XYL	Wife
YL	Young lady
73	Best regards
88	Love and kisses

Resources for Shortwave Listeners

Clubs for Shortwave Listeners

American Shortwave Listeners Club, 16182 Ballad Lane, Huntington Beach, CA, 92649. All-band coverage with a monthly bulletin. This club has been around for over two decades but has had some problems with irregular bulletins lately.

Association of Clandestine Radio Enthusiasts, P. O. Box 11201, Shawnee Mission, KS, 66207-0201. Exclusively devoted to clandestine radio, pirate stations, numbers stations, and other unexplained and mysterious radio through its monthly bulletin. For my money, this is the most interesting radio club around.

Association of DX Reporters, 7008 Plymouth Rd., Baltimore, MD, 21208. All-band coverage with monthly bulletin. Most members are on the east coast.

Fine Tuning, P. O. Box 780075, Wichita, KS, 67278. Not really a club, but more of a loose confederation of those interested in chasing difficult SW broadcast DX. Weekly bulletin during fall and winter. Best coverage of SW DX anywhere.

International Radio Club of America, 159 Old Post Road North, Croton-on-Hudson, NY, 10520. Covers BCB DX only; bulletin is published weekly during the winter DX season. Many useful technical and feature articles.

Longwave Club of America, 45 Wildflower Rd., Levittown, PA, 19057. Exclusive coverage of longwave DX through its monthly bulletin.

National Radio Club, P. O. Box 164, Mannsville, NY, 13661. Covers BCB DX only, and has been around since the 1930s. Bulletin is published weekly during the winter. Superb technical articles and coverage of foreign BCB DX.

North American Shortwave Association, 45 Wildflower Rd., Levittown, PA, 19057. Covers SW broadcast only through it monthly bulletin. Currently, this is the best bulletin of any club; membership is essential if you're interested in SW broadcasting.

Commercial SWL Publications

Clandestine Confidential Newsletter, RR #4, Box 110, Lake Geneva, WI, 53147. Bi-monthly roundup of latest clandestine radio developments under the editorship of well-known DXer Gerry Dexter.

CRB Research, P. O. Box 56, Commack, NY, 11725. The most extensive selection of SWLing and radio-related books anywhere. If they don't have it, it doesn't exist.

MoonBeam Press, P. O. Box 149 Briarcliff, NY, 10510. Publisher of several newsletters for SWLs and DXers.

Monitoring Times, 140 Dog Branch Rd., Brasstown, NC, 28902. Covers a wide range of communications topics, with an emphasis on utilities and VHF/UHF scanning channels. Good coverage of RTTY reception techniques.

Popular Communications, 76 North Broadway, Hicksville, NY, 11801. If you're interested in the topics covered in this book, you must read this magazine. Monthly columns on shortwave broadcasting, RTTY, clandestine radio, and pirate stations along with numerous feature articles.

Books for Shortwave Listeners

Los Numeros: A Guide to the Numbers Stations, Tiare Publications, P. O. Box 493, Lake Geneva, WI, 53147. A thorough discussion of numbers stations written by a former intelligence officer who goes by the pseudonym of "Havana Moon."

Passport to World Band Radio, P. O. Box 300, Penn's Park, PA, 18943. Computer-generated listing of SW stations by frequency with complete schedules, languages used, and other information. Several feature articles and superb equipment reviews. Issued annually; an essential reference for any SWL.

Pirate Radio Directory, Tiare Publications, P. O. Box 493, Lake Geneva, WI, 53147. George Zeller's annual guide to pirate radio stations is the most complete and interesting book on the subject; even includes addresses for most pirates!

Shortwave Directory, Grove Enterprises, 140 Dog Branch Road, Brasstown, NC, 28902. Comprehensive directory of utility stations by function (aeronautical, maritime, etc.) and by operating agency. A good basic reference work for utility DXers.

Top Secret Registry of U.S. Government Frequencies, CRB Research, P. O. Box 56, Commack, NY, 11725. Covers frequencies used by various government agencies, although none (despite title) are actually classified information. Serious utility listeners will need this one.

World Radio Television Handbook, Billboard Publications, 1515 Broadway, New York, NY, 10036. Annual directory of shortwave stations in a by-country format with a few feature articles; another important reference for serious SWLs.

Index